Maintaining U.S. Leadership in Aeronautics

Scenario-Based Strategic Planning for NASA's
Aeronautics Enterprise

Steering Committee for a Workshop to Develop
Long-Term Global Aeronautics Scenarios

Aeronautics and Space Engineering Board

Commission on Engineering and Technical Systems

National Research Council

NATIONAL ACADEMY PRESS
Washington, D.C. 1997

NOTICE: The project that is the subject of this report was approved by the Governing Board of the National Research Council, whose members are drawn from the councils of the National Academy of Sciences, the National Academy of Engineering, and the Institute of Medicine. The members of the committee responsible for the report were chosen for their special competencies and with regard for appropriate balance.

This report has been reviewed by a group other than the authors according to procedures approved by a Report Review Committee consisting of members of the National Academy of Sciences, the National Academy of Engineering, and the Institute of Medicine.

The National Academy of Sciences is a private, nonprofit, self-perpetuating society of distinguished scholars engaged in scientific and engineering research, dedicated to the furtherance of science and technology and to their use for the general welfare. Upon the authority of the charter granted to it by the Congress in 1863, the Academy has a mandate that requires it to advise the federal government on scientific and technical matters. Dr. Bruce M. Alberts is president of the National Academy of Sciences.

The National Academy of Engineering was established in 1964, under the charter of the National Academy of Sciences, as a parallel organization of outstanding engineers. It is autonomous in its administration and in the selection of its members, sharing with the National Academy of Sciences the responsibility for advising the federal government. The National Academy of Engineering also sponsors engineering programs aimed at meeting national needs, encourages education and research, and recognizes the superior achievements of engineers. Dr. William A. Wulf is interim president of the National Academy of Engineering.

The Institute of Medicine was established in 1970 by the National Academy of Sciences to secure the services of eminent members of appropriate professions in the examination of policy matters pertaining to the health of the public. The Institute acts under the responsibility given to the National Academy of Sciences by its congressional charter to be an adviser to the federal government and, upon its own initiative, to identify issues of medical care, research, and education. Dr. Kenneth I. Shine is president of the Institute of Medicine.

The National Research Council was organized by the National Academy of Sciences in 1916 to associate the broad community of science and technology with the Academy's purposes of furthering knowledge and advising the federal government. Functioning in accordance with general policies determined by the Academy, the Council has become the principal operating agency of both the National Academy of Sciences and the National Academy of Engineering in providing services to the government, the public, and the scientific and engineering communities. The Council is administered jointly by both Academies and the Institute of Medicine. Dr. Bruce M. Alberts and Dr. William A. Wulf are chairman and interim vice-chairman, respectively, of the National Research Council.

This study was supported by Contract No. NASW-4938 between the National Academy of Sciences and the National Aeronautics and Space Administration. Any opinions, findings, conclusions, or recommendations expressed in this publication are those of the author(s) and do not necessarily reflect the view of the organizations or agencies that provided support for the project.

International Standard Book Number: 0-309-05696-9

Available in limited supply from
Aeronautics and Space Engineering Board
2101 Constitution Avenue, N.W.
Washington, DC 20418
(202) 334-2855

Copyright 1997 by the National Academy of Sciences. All rights reserved.

Printed in the United States of America.

Steering Committee for a Workshop to Develop Long-Term Global Aeronautics Scenarios

WILLIAM W. HOOVER (chair), U.S. Air Force (retired), Williamsburg, Virginia
GUION S. BLUFORD, NYMA, Inc., Brook Park, Ohio
RICHARD S. GOLASZEWSKI, GRA, Inc., Jenkintown, Pennsylvania
WILLIAM H. HEISER, U.S. Air Force Academy, USAF Academy, Colorado
GRACE M. ROBERTSON, McDonnell Douglas Corporation (Douglas Aircraft Company), Long Beach, California
JEFFREY K. SCHWEITZER, United Technologies/Pratt & Whitney, West Palm Beach, Florida.
THOMAS B. SHERIDAN, Massachusetts Institute of Technology, Cambridge
ROBERT E. SPITZER, Boeing Commerical Airplane Group, Seattle, Washington

Aeronautics And Space Engineering Board Staff

David A. Turner, Study Director
JoAnn Clayton-Townsend, ASEB Director
Victoria P. Friedensen, Senior Project Assistant
Ted Morrison, Senior Project Assistant

Transportation Research Board Staff Liaison

Joseph Breen, Aviation Specialist

Aeronautics and Space Engineering Board

JOHN D. WARNER (chair), The Boeing Company, Seattle, Washington
STEVEN AFTERGOOD, Federation of American Scientists, Washington, D.C.
GEORGE A. BEKEY, University of Southern California, Los Angeles
GUION S. BLUFORD, JR., NYMA, Inc., Brook Park, Ohio
RAYMOND S. COLLADAY, Lockheed-Martin Astronautics, Denver, Colorado
BARBARA C. CORN, BC Consulting, Inc., Searcy, Arizona
STEVEN D. DORFMAN, Hughes Electronics Corporation, Los Angeles, California
DONALD C. FRASER, Boston University, Boston, Massachusetts
DANIEL HASTINGS, Massachusetts Institute of Technology, Cambridge
FREDERICK HAUCK, International Technology Underwriters, Bethesda, Maryland
WILLIAM H. HEISER, U.S. Air Force Academy, USAF Academy, Colorado
WILLIAM W. HOOVER, U.S. Air Force (retired), Williamsburg, Virginia
BENJAMIN HUBERMAN, Huberman Consulting Group, Washington, D.C.
BERNARD L. KOFF, Pratt & Whitney, West Palm Beach, Florida
FRANK E. MARBLE, California Institute of Technology, Pasadena
C. JULIAN MAY, Technical Operations International, Inc., Kennesaw, Georgia
GRACE M. ROBERTSON, McDonnell Douglas Corporation (Douglas Aircraft Company), Long Beach, California
GEORGE SPRINGER, Stanford University, Stanford, California

Staff

JoAnn Clayton-Townsend, Director

Preface

THE TASK

In June 1996, the Office of Aeronautics of the National Aeronautics and Space Administration (NASA) approached the National Research Council (NRC) with a request to conduct a workshop to help guide their strategic planning process.[1] At the workshop, five future scenarios of the world, based primarily on economic, social, and policy factors, would be refined and further developed by considering the potential for revolutionary and evolutionary technology developments, by determining the key issues for aeronautics, and by discussing the general role that NASA should play in addressing the key issues in the future. These scenarios would span the breadth of aeronautics, including civil aviation, military aviation, and access to space.[2] In addition, the scenarios would consider other critical factors that could impact the future of aeronautics, including information and communications systems, national and global transportation systems, and air traffic management systems. Long-term was defined as 15 to 25 years or beyond the next-generation systems.

The pre-workshop development of the scenarios would be accomplished by a core team of individuals from the NASA Office of Aeronautics, working in collaboration with The Futures Group (TFG), a firm specializing in corporate strategic planning, and the Systems Technology Group of Science Applications International Corporation (SAIC).

STUDY APPROACH

A steering committee was formed under the auspices of the Aeronautics and Space Engineering Board to accomplish the task of planning and conducting the workshop, advising on the pre-workshop selection of the long-term global scenarios, and developing a report that would convey the results of the workshop. Appendix B contains brief biographies of the steering committee members. Before the workshop the steering committee met on two occasions, on August 7 and September 16, 1996, and after the workshop on October 3 and October 28, 1996, to finalize this report. The workshop itself took place on September 30 to October

---

[1] The statement of task, as approved by the governing board of the NRC, can be found in Appendix A.

[2] In addressing the "access to space" area of aeronautics research and development, the steering committee, the scenarios, and the workshop addressed only potential requirements for launch vehicles for unmanned Earth-orbital space applications. Solar and deep space exploration and the manned space program were not considered.

2, 1996. In consultation with the study's sponsor, the term "key issues" was defined further by the steering committee as key needs and opportunities for aeronautics and the technological implications of these needs and opportunities.

During the pre-workshop scenario development process most steering committee members were interviewed by a member of the NASA/TFG/SAIC core team. The steering committee also participated in the scenario selection process that determined which five of sixteen possible scenarios would be presented and discussed at the workshop.[3] The specific narratives for each scenario in Appendix D, however, were developed solely by the core team.

At the workshop, the steering committee employed the scenario-based strategic planning process described in Chapter 1 and in Appendix H, led each of the "world teams" in their analysis of the scenarios, and fostered a cohesive and interactive exchange of ideas among the widely diverse and knowledgeable group of participants.

ORGANIZATION OF THE REPORT

This report begins with an Executive Summary that outlines the results of this study based on the workshop and the deliberations of the steering committee. In Chapter 1 the need for strategic planning in aeronautics is discussed, and the pre-workshop scenario development process and the workshop agenda are briefly described. Each of the five future scenarios that were examined during the workshop are summarized in Chapter 2. Each summary includes a discussion of the role of aviation and aeronautics in the scenario and a list of the needs and opportunities for aeronautics as determined by the working group that focused on the scenario during the workshop. In Chapter 3 the collective results of the workshop and the additional deliberations of the steering committee are presented. The general role that the steering committee believes NASA should play in maintaining the superiority of U.S. aeronautics products and services in the future also is discussed. Several appendices are included that provide additional information on various aspects of the workshop and the study in general. Notes in the report's main text refer the reader to a related appendix when appropriate.

ACKNOWLEDGMENTS

The steering committee thanks all those who participated in the workshop for their extraordinary contributions and sense of commitment to the task.

---

[3] A complete list of the workshop participants can be found in Appendix C.

Contents

EXECUTIVE SUMMARY 1

1 INTRODUCTION 9
      Need for Strategic Planning in Aeronautics, 9
      Pre-Workshop Development of the Scenarios, 10
      The Workshop, 14
      Maintaining U.S. Competitiveness in Aeronautics, 14
      References, 15

2 LONG-TERM GLOBAL SCENARIOS AND THEIR IMPLICATIONS FOR
AERONAUTICS 17
      Introduction, 17
      Pushing the Envelope, 17
      Grounded, 22
      Regional Tensions, 25
      Trading Places, 30
      Environmentally Challenged, 33
      Summary of Needs and Opportunities, 37

3 NEEDS AND OPPORTUNITIES, TECHNOLOGY IMPLICATIONS, AND THE
FUTURE ROLE OF NASA 41
      Future Needs and Opportunities, 41
      System Level Technology Implications, 44
      NASA's Future Role in Maintaining U.S.
        Competitiveness in Aeronautics, 47
      References, 50

APPENDICES

    A   Statement of Task 53
    B   Biographical Sketches of Steering Committee Members 55
    C   Workshop Participants 57
    D   Scenario Narratives as Provided by the
         NASA/TFG/SAIC Core Team 61
    E   Workshop Agenda 117
    F   Questions for World Team Sessions 121
    G   Bibliography 123
    H   Scenario-Based Strategic Planning as
         Described by the Futures Group 127

List of Tables and Figures

TABLES

ES-1   The Five Scenarios and Their Dimensions, 2
ES-2   Robust, Significant, and Noteworthy Needs and Opportunities, 3
2-1   Summary of Needs and Opportunities for Each Scenario, 38
3-1   Robust, Significant, and Noteworthy Needs and Opportunities, 42

FIGURES

1-1   The 16 possible scenarios based on four dimensions, 12
1-2   The pre-workshop scenario development process, 13
H-1   NASA global scenarios, 128
H-2   Organizational strategies, 129

Executive Summary

The Office of Aeronautics of the National Aeronautics and Space Administration (NASA) asked the National Research Council (NRC) to conduct a workshop to help guide their strategic planning process. As part of this process, but apart from the NRC effort, a core team of individuals from the NASA Office of Aeronautics, The Futures Group (TFG), and Science Applications International Corporation (SAIC), developed five long-term global aeronautics scenarios, where long-term was defined as 15 to 25 years hence or beyond the next-generation systems. These scenarios are based primarily on economic, social, and policy issues identified by experts and leaders in the field of aeronautics during interviews conducted by the core team.[1] These issues or "drivers" were woven into the narratives of each of the five scenarios and were used to determine the general characteristics or "dimensions" of each scenario. The names given to these five scenarios, and their dimensions as defined by U.S. economic competitiveness, worldwide demand for aeronautics products and services, threats to global security and/or quality of life, and the global trend in government participation in society, are summarized in Table ES-1.

The workshop was planned and conducted by a steering committee formed under the auspices of the Aeronautics and Space Engineering Board. Workshop participants from aeronautics- and aviation-related organizations in the public and private sectors were asked to accept the generalities of each scenario as provided and to further refine and develop each scenario by (1) considering potential revolutionary and evolutionary technology developments, (2) determining the key needs and opportunities for aeronautics and related technology implications, and (3) discussing the general role that NASA should play in maintaining the future competitiveness of U.S. aeronautics.

NEEDS AND OPPORTUNITIES FOR AERONAUTICS

Five working groups or "world teams" were formed and led by the steering committee to accomplish the task outlined above. Each team focused on one scenario for the majority of the three-day workshop, but an iterative round-robin process was also utilized to determine the applicability of needs and opportunities for

---

[1] Experts and leaders interviewed by members of the core team included members of the ASEB steering committee and many of the workshop participants. The steering committee also participated in the process of defining the dimensions of the scenarios and helped to choose the scenarios that would be considered at the workshop. However, the narratives of the scenarios were developed solely by the core team.

aeronautics that were developed by other teams for their own world. By this process, needs and opportunities were found that were robust (cross-cutting items that fit within the environment of each future scenario), significant (items that were critically important to three or four scenarios), and noteworthy (items that were novel and were important to two scenarios or less). These needs and opportunities are summarized in Table ES-2.

TABLE ES-1 The Five Scenarios and Their Dimensions

Scenario	U.S. Economic Competitiveness[a]	Worldwide Demand for Aeronautics Products and Services[b]	Threats to Global Security and/or Quality of Life[c]	Global Trend in Government Participation in Society[d]
Pushing the Envelope	Strong	High growth	Low	Low
Grounded	Strong	Low growth	High	High
Regional Tensions	Weak	High growth	High	High
Trading Places	Weak	High growth	Low	Low
Environmentally Challenged	Weak	Low growth	High	High

[a] The relative U.S. share of internationally traded products and services in the world economy (strong or weak).
[b] The level of demand for aeronautics products and services related to civil; military; and access to space applications in local, regional, and global markets (high growth and low growth).
[c] Direct threats to the health and safety of people, and/or the stability and viability of governments, and their implications for the United States (high or low threat).
[d] The tendency of governments to regulate and/or intervene in key aspects of society and the economy (high or low).

SYSTEM LEVEL TECHNOLOGY IMPLICATIONS

The steering committee's synthesis of the needs and opportunities discussed at the workshop, and their implications for broad areas of technology development, is provided below. The system level technologies identified do not appear in order of priority and have not been analyzed comprehensively to determine their relative scientific merit or technical feasibility. They simply represent the principal items discussed based on an analysis of the five scenarios and the iterative round-robin process. Further analysis will be needed in order to justify spending scarce research and development (R&D) funds on many of these areas of technology.

New Aircraft

New markets are likely to demand new types of aircraft, including short-range cargo and passenger aircraft; long-range, high-capacity supersonic aircraft; modular and reconfigureable aircraft; aircraft with autonomous, on-board air traffic control capability; aircraft with built-in redundancy and self-inspection to simplify repair and maintenance; and aircraft that utilize smart structures and tailored materials.

TABLE ES-2 Robust, Significant, and Noteworthy Needs and Opportunities

ROBUST Common to all scenarios	SIGNIFICANT Less common but vital to some scenarios	NOTEWORTHY Specialized and unique
Air Traffic Management satellite-based, autonomous, tailored	Access to Space small payloads, low cost, on demand	Short-to-Medium Range Aircraft VSTOL, commuter, infrastructure independent, military special operations
Airport Infrastructure constrained, austere, tailored	Supersonic Aircraft long range, large, and low capacity	Stealth Aircraft evade terrorist threats, quiet over populated areas
Safety/ Survivability significant accident reduction, survive natural and man-made threats	Subsonic Aircraft large, small, long and short range Air Cargo large, low-cost, specialized and reconfigurable aircraft	General Aviation increased activity, part of a customer-tailored air transportation system
Manufacturing agile, virtual, validation, certification	Uninhabited Air Vehicles weapons, surveillance, intelligence	Tailored and Smart Materials reduced fuel consumption and enhanced safety
	Environment noise, emissions, hydrogen fuels	Microelectro Mechanical Systems reduced fuel consumption and vehicle size
	Security Systems airport, aircraft, terrorist threat	Sonic Boom Mitigation enable supersonic flight over populated areas
	Vertical/Short Takeoff and Landing (VSTOL) Aircraft short, medium, and long range, stealth, infrastructure independent, military special operations	
	Skilled Training and Education distributed and tailored training	

System Integration in Aircraft Design, Manufacturing, and Operations

To maintain its competitive position, the United States will need to foster improved modeling, virtual reality,[2] and planning tools for agile and flexible manufacturing.

---

[2] Virtual reality is a computer-based technology that allows the user to interact with data that give the appearance of a three-dimensional environment or world. The user can "enter" and "navigate" the three-dimensional world portrayed as graphic images and interact with objects in that world as if "inside" that world.

The integration of avionics and other information systems within aircraft, within the air traffic control system, and within other portions of the air transportation system will need to be undertaken to ensure reliable communications and efficiency of operation.

Passenger and Crew Safety and Security

Increasing demand for operational safety, combined with public awareness of the threat of terrorism and the growth in the volume of air traffic will require improved aviation weather observation dissemination to end users; decision aids on board aircraft to mitigate human error; improved aircraft system reliability through fault tolerance and artificial intelligence; improved systems within aircraft to detect explosives and contraband; and improved aircraft survivability to weapons, electromagnetic impulses, radio frequency interference, and severe weather phenomena.

Improved Operating Efficiency and Cost Effectiveness

The United States can improve the cost effectiveness and operating efficiency of both civil and military aviation by pioneering the following capabilities: increased automation of aircraft control and air traffic management, including autonomous operations both in the air and on taxiways; uninhabited air vehicles (UAVs) to be used initially for surveillance and weather observation and eventually for aerial combat and cargo transport; tailored materials (designed at the molecular level) and smart materials (able to sense their own conditions) to increase safety, reduce aircraft weight, and to withstand higher engine temperatures; high-efficiency subsonic propulsion systems that provide improved fuel consumption; and finally, miniaturized electronics, sensors, and other nonstructural mechanical components to reduce weight and enable new aircraft and propulsion system designs.

Environmental Protection and Noise Abatement

The public will continue to demand reductions in environmental damage and reductions of acoustic noise over urban areas. This will require the United States to collaborate with other nations to develop technology that will reduce or eliminate harmful aircraft engine emissions and technology that will enable quieter engines and operations, including revolutionary means to mitigate sonic boom effects over populated areas.

Access to Space

Earth orbit will continue to provide opportunities for both civilian and military use of satellites for communications, navigation, and surveillance. It is anticipated that commercial firms and other nations will want on-demand access to Earth orbit that is quick and inexpensive. In most cases, sensors and communication devices are likely to be the dominant payloads, and there will be no inherent need for large spacecraft. Therefore, the steering committee has identified a need for systems and associated infrastructures to enable low-cost, on-demand delivery of small payloads with sensors or communications packages to Earth orbit. Potential needs for future manned space activities and space science missions were not discussed by the steering committee or the workshop participants.

NASA'S FUTURE ROLE IN MAINTAINING U.S. COMPETITIVENESS IN AERONAUTICS

The workshop represented a microcosm of a real partnership between government, industry, and academia that must continue to be fostered to achieve the goals outlined by the National Science and Technology Council (NSTC) and therefore enable a competitive U.S. aeronautics industry in the future. Within this partnership, government must ensure that the conduct of basic and applied research, the development of high-risk technology, the rapid validation of essential design and manufacturing tools and techniques, and the certification of products continue to be focused on these goals. The steering committee believes that within the federal government, coordinated, cost-effective planning and implementation of long-term aeronautics R&D can only be accomplished by using the interagency process to designate a lead agency for this role. The steering committee further believes that NASA would best serve as the lead agency, rather than the U.S. Department of Defense (DOD), the Federal Aviation Administration (FAA), the National Science Foundation, or the National Institute of Standards and Technology for the following reasons.[3]

- NASA is chartered by the National Aeronautics and Space Act of 1958 to "preserve the role of the United States as a leader in aeronautical science and technology and the application thereof." No other federal agency has this legislative mandate.
- The NASA aeronautics enterprise has inherited its fundamental aeronautics R&D focus from its forerunner, the National Advisory Committee for

---

[3] A session also was held at the workshop to discuss possible options for the future conduct of aeronautics R&D and the proper role of NASA in the future. NASA center directors and members of the NASA/TFG/SAIC core team were excluded from this session. The steering committee's view on NASA's future role in maintaining the superiority of U.S. aeronautics products and services is based, in part, on the deliberations that took place during this session.

Aeronautics, chartered in 1915. NASA has maintained this focus and has developed and maintained extensive R&D equipment and facilities. Although other federal agencies, such as the DOD and the FAA, also conduct aeronautics R&D and maintain appropriate facilities, this work is carried out in support of their operational missions. In contrast, the mission of NASA's aeronautics enterprise *is* aeronautics R&D. NASA has been charged by the Office of Management and Budget to develop an integrated national strategy and priorities assessment for civil aeronautics (NASA, 1995).

- NASA has responded to the goals outlined by the NSTC through its *Aeronautics Strategic Enterprise Plan for 1995–2000* (NASA, 1995). This plan includes a preliminary "road map," or strategic plan, for the future of aviation that will be refined as a result of the workshop, this report, and the larger strategic planning process currently under way in the NASA Office of Aeronautics.
- NASA has several programs currently under way that already involve substantial partnerships between government, industry, and academia. These programs include the Advanced Subsonic Technology (AST) program, the High Speed Research (HSR) program, and the Advanced General Aviation Technology (AGATE) program.
- The future needs, opportunities, and implications for technology discussed in this report offered no compelling reason for the workshop participants or the steering committee to recommend an alternative to future NASA leadership, although the alternatives mentioned previously were considered.

Recommendation. To ensure coordinated, cost-effective planning and implementation of long-term aeronautics research and development within the federal government, the interagency process should be used to designate a lead agency for this role. The steering committee believes that NASA would best serve as the lead agency.

Leadership does not imply that NASA will lead every R&D activity that is focused on future aircraft, systems, and technology areas recommended for future development by this report.[4] Therefore, an in-depth assessment of the specific programs and long-term R&D activities that NASA should engage in as the lead agency for aeronautics is the next logical step in this current strategic planning process. In addition, the roles of other federal agencies, private sector organizations, and academic institutions that are part of the nation's aeronautics partnership must be carefully considered and defined. The steering committee believes that this next phase of the strategic planning process should again be conducted with broad participation from government, industry, and academia and should proceed without delay.

---

[4] The steering committee envisions that NASA's role in the development of technology would not extend beyond what is referred to by the DOD as 6.3A—Advanced Development.

REFERENCE

NASA (National Aeronautics and Space Administration). 1995. *Achieving Aeronautics Leadership: Aeronautics Strategic Enterprise Plan, 1995–2000.* Washington, D.C.: National Aeronautics and Space Administration.

1

Introduction

NEED FOR STRATEGIC PLANNING IN AERONAUTICS

As the Aeronautics and Space Engineering Board (ASEB) stated in *Aeronautical Technologies for the Twenty-First Century*, the U.S. aeronautics industry has been one of the undisputed success stories in global competitiveness throughout the latter half of this century. Since the end of World War II, the United States has been a leader in the global aeronautics industry, and, in most cases, U.S. aircraft, engines, and parts have dominated both domestic and foreign markets for subsonic transports, general aviation, commuter, and military aircraft. The buildup of the global transportation infrastructure (i.e., airports and air traffic management systems) has also been driven by U.S. technology and products. The aeronautics industry, one of the largest positive industrial contributors to the U.S. balance of trade, plays a vital role in maintaining the safety and convenience of air travel throughout the world and provides important contributions to the defense of U.S. interests (NRC, 1992).

However, as of 1992 the U.S. market share in aeronautics had eroded as a result of foreign competitors that brought products to market that have lower total ownership costs than U.S. products (GRA, 1990). Lower total ownership costs can be achieved, for example, through implementation of new technologies that reduce long-term operating costs or through products that enter the market with significantly lower purchase prices. The ASEB report, *Aeronautical Technologies for the Twenty-First Century*, took issue with a common misconception that aeronautics is a mature industry. Many areas where significant technical progress remains to be made were identified in the report (NRC, 1992). Although the erosion in market share seems to have leveled off in the last few years, it seems clear that advances in technology will continue to be a significant element in maintaining U.S. market leadership and economic competitiveness in the future.

As part of a study completed in 1994, the National Research Council (NRC) Committee on Japan developed future scenarios for the course of the global aircraft industry and United States–Japan alliances over the next decade and beyond. The Committee on Japan concluded that several scenarios contemplating a decline in U.S. market share for aeronautics products and services were plausible if current trends in the aerospace industry continue. Furthermore, the committee concluded that for the United States to maintain its leadership in this critically important industrial sector, government-industry partnerships for the development and implementation of a long-term strategy are essential (NRC, 1994).

Efforts have been under way within the federal government to address the concerns of these two NRC reports that discuss the competitiveness of the nation's aeronautics research and development (R&D) enterprise. In 1995 the National Science and Technology Council (NSTC) released a report that acknowledged the need for action to ensure that the United States maintains a strong and competitive aeronautics industry. Three goals were identified that could be accomplished through a government, industry, and university partnership in aeronautics research and technology development (NSTC, 1995):

- maintain the superiority of U.S. aircraft and engines[1]
- improve the safety, efficiency, and cost effectiveness of the global air transportation system
- ensure the long-term environmental compatibility of the aviation system

Within the National Aeronautics and Space Administration (NASA), which is chartered by the National Aeronautics and Space Act of 1958 in part to "preserve the role of the United States as a leader in aeronautical science and technology and the application thereof," the Office of Aeronautics has developed a strategic plan for 1995 to 2000 (NASA, 1995) that follows many of the programmatic recommendations in *Aeronautical Technologies for the Twenty-First Century* (NRC, 1992) and supports the goals outlined by the NSTC. This plan even includes an attempt to characterize aviation in 2020 and lists several types of aerospace systems and technologies that will likely be in use. However, this is only a preliminary vision based on the judgment of NASA aeronautical experts and the extrapolation of current trends in aviation and aeronautics.

PRE-WORKSHOP DEVELOPMENT OF THE SCENARIOS

Recognizing that a long-term strategic plan for aeronautics requires a broad-based national perspective that includes the needs of users and consumers, the NASA Office of Aeronautics asked the NRC to conduct a workshop that would bring together experts from industry, government, and academia to analyze a number of possible scenarios for aeronautics 15 to 25 years hence. A steering committee was formed under the auspices of the ASEB to plan, organize, and conduct the workshop and report on its conclusions. However, the pre-workshop assignment to develop future

---

[1] For the purposes of this study, the steering committee rephrased this goal as "maintain the superiority of U.S. aeronautics products and services." The steering committee believes that this appropriately broadens the goal to include subsystems of aircraft and engines, manufacturing processes related to the production of aircraft and engines, and research and engineering services related to aeronautics.

scenarios based primarily on economic, social, and policy factors was performed by a core team of individuals from the NASA Office of Aeronautics, working in collaboration with The Futures Group (TFG), and the Systems Technology Group of Science Applications International Corporation (SAIC). The NASA/TFG/SAIC team, hereafter referred to as the core team, began the process of developing these scenarios by conducting a number of interviews with experts and leaders in the field of aeronautics, including members of the ASEB steering committee and many of the workshop participants. Through a series of questions, the interviewees were asked to identify key issues related to the air transportation system and the worldwide aeronautics industry. They also were asked to suggest critical factors or "drivers" that would affect the future of these industries.

These drivers were then used to define a future operating environment that was as inclusive as possible.[2] This environment was bound by a subset of the drivers that became the "dimensions" of the operating environment or "scenario space." These dimensions, which were defined by the core team and the steering committee, are as follows:

- U.S. economic competitiveness—the relative U.S. share of internationally traded products and services in the world economy (strong or weak)
- Worldwide demand for aeronautics products and services—the level of demand for aeronautics products and services related to civil, military, and access to space applications in local, regional, and global markets (high growth and low growth)
- Threats to global security and/or quality of life—direct threats to the health and safety of people, and/or the stability and viability of governments, and their implications for the United States (high or low threat)
- Global trend in government participation in society—the tendency of governments to regulate and/or intervene in key aspects of society and the economy (high or low)

Altering four dimensions that each have two values creates a total of 16 possible scenarios. These 16 possibilities are illustrated in Figure 1-1. Five of the scenarios were selected for further analysis at the workshop by the steering committee, the core team, and the senior leaders in NASA's aeronautics enterprise based on the potential challenges or opportunities they may hold for aeronautics. These five were given the following titles: Pushing the Envelope, Grounded, Regional Tensions, Trading Places, and Environmentally Challenged. It is important to note that the workshop participants were asked to accept each scenario as provided rather than challenge their content.

---

[2] A summary of the drivers mentioned during the interviews and the role that each driver played in the development of the scenarios is provided at the end of each scenario narrative in Appendix D.

	U.S. Economic Competitiveness		Worldwide Demand for Aeronautics Products and Services		Threats to Global Security and/or Quality of Life		Global Trend in Government Participation in Society		
	Strong	Weak	High Growth	Low Growth	High	Low	Low	High	
1	x		x		x		x		
2	x		x		x			x	
3	x		x			x	x		Pushing the Envelope
4	x		x			x		x	
5	x			x	x		x		
6	x			x	x			x	Grounded
7	x			x		x	x		
8	x			x		x		x	
9		x	x		x		x		
10		x	x		x			x	Regional Tensions
11		x	x			x	x		Trading Places
12		x	x			x		x	
13		x		x	x		x		
14		x		x	x			x	Environmentally Challenged
15		x		x		x	x		
16		x		x		x		x	

FIGURE 1-1 The 16 possible scenarios based on four dimensions.

The final step in the pre-workshop scenario development process, providing a descriptive narrative to each of the five selected scenarios that was both plausible and internally consistent, was accomplished by the core team. Figure 1-2 depicts this complete process.

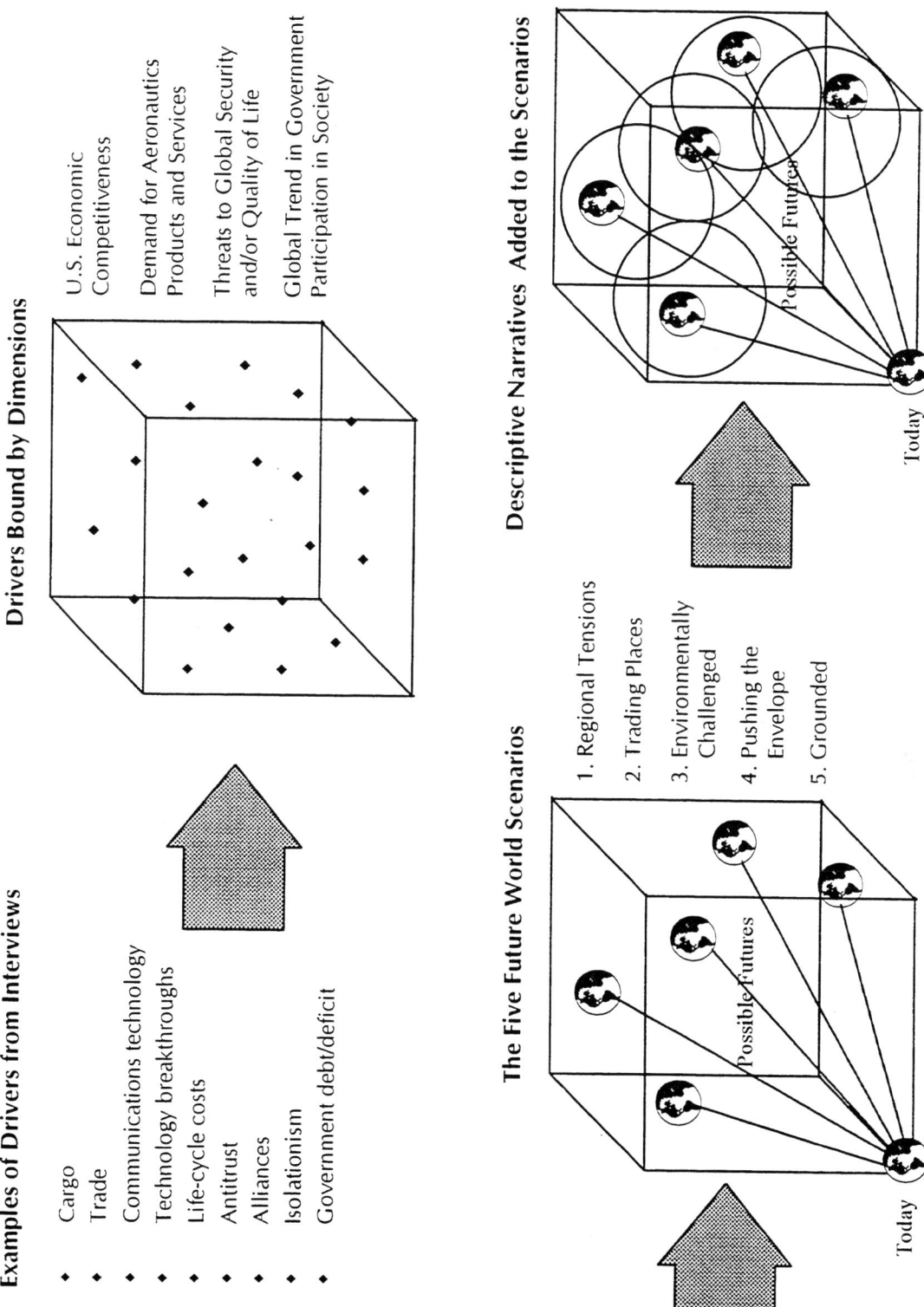

FIGURE 1-2 The pre-workshop scenario development process.

THE WORKSHOP

The workshop, planned and organized by the ASEB steering committee, took place on September 30 to October 2, 1996 (see Appendix E). Participants were divided into five working groups or "world teams" that were each led by a steering committee member. Each group focused on one scenario for the majority of the three days. Each answered a series of questions prepared jointly by the steering committee and the core team that were designed to help the team "live in its world" in order to develop a comprehensive view of the role that aeronautics would play in the scenario (see Appendix F). Once this role was defined, needs, opportunities, and their implications for aeronautics were determined for the team's scenario.

After the world teams completed this process for their scenarios, each team participated in a round-robin process to determine the applicability of other teams' needs and opportunities for aeronautics to their own scenario. This iterative process also helped each team benefit from unique expertise residing on other teams that would not have been available to them otherwise. Concepts, strategies, and technologies that matched needs and opportunities in a team's own world were added to a final list presented by each team leader in the plenary session on the afternoon of the final day of the workshop.

An executive session also was held at the workshop to discuss possible options for the future conduct of aeronautics R&D and to discuss the proper role of NASA in the future. NASA Center Directors and members of the NASA/TFG/SAIC core team were excluded from this session. The steering committee's view on NASA's future role in maintaining the superiority of U.S. aeronautics products and services is based in part on the deliberations that took place during this session.

MAINTAINING U.S. COMPETITIVENESS IN AERONAUTICS

The steering committee and participants agreed early in the workshop that it is essential for the United States to maintain its superiority in aeronautics products and services by continually improving safety, efficiency, cost, and long-term environmental compatibility. Without leadership, effective strategic planning and subsequent R&D implementation, aeronautics technology and jobs could quickly move to other nations, adversely affecting employment opportunities, trade balances, national security, and the efficiency of our transportation system. Those in attendance at the workshop also agreed that global competitiveness can only be maintained through a strong government, industry, and academia partnership. The NSTC reached a similar conclusion in its 1995 report, *Goals for a National Partnership in Aeronautics Research and Technology* (NSTC, 1995), as did the Council on Competitiveness in *Endless Frontier, Limited Resources—U.S. R&D Policy for Competitiveness* (Council on Competitiveness, 1996).

The process of defining, developing, and deliberating over future world scenarios as a means of long-range strategic planning is most often applied by organizations in both the public and private sectors that have a single-minded purpose (i.e., profit and customer satisfaction) or a singular mission (e.g., protect the borders of the United States). In the case of this workshop, however, a purposely diverse group of experts from government, industry, and academia was asked to participate in the process. For this reason, the members of the steering committee believed that it would be helpful to instill all participants with a common sense of purpose by creating a "virtual" corporation or agency during the three-day workshop. This virtual corporation was called "U.S. Aeronautics, Inc."

Once the workshop participants assumed the identity of U.S. Aeronautics, Inc., the group engaged in a strategic planning process that was focused on accomplishing the three goals defined by the NSTC, regardless of what the future may hold in 15 to 25 years.

REFERENCES

Council on Competitiveness. 1996. Endless Frontier, Limited Resources—U.S R&D Policy for Competitiveness. Washington, D.C.: Council on Competitiveness.

GRA (Gellman Research Associates). 1990. An Economic and Financial Review of Airbus Industrie. Prepared for the U.S. Department of Commerce International Trade Administration. Jenkintown, Pa.: Gellman Research Associates.

NASA (National Aeronautics and Space Administration). 1995. Achieving Aeronautics Leadership—Aeronautics Strategic Enterprise Plan, 1995–2000. Washington, D.C.: National Aeronautics and Space Administration.

NRC (National Research Council). 1992. Aeronautical Technologies for the Twenty-First Century. Aeronautics and Space Engineering Board, Committee on Aeronautical Technologies. Washington, D.C.: National Academy Press.

NRC. 1994. High Stakes Aviation: U.S.-Japan Technology Linkages in Transport Aircraft. Office of Japan Affairs, Committee on Japan. Washington, D.C.: National Academy Press.

NSTC (National Science and Technology Council). 1995. Goals for a National Partnership in Aeronautics Research and Technology. Executive Office of the President, Office of Science and Technology Policy. Washington, D.C.: National Science and Technology Council.

2
Long-Term Global Scenarios and Their Implications for Aeronautics

INTRODUCTION

Each of the five future scenarios that were considered at the workshop is discussed in separate sections of this chapter. They are entitled "Pushing the Envelope," "Grounded," "Regional Tensions," "Trading Places," and "Environmentally Challenged." Each section contains a short summary of the scenario as it was provided to the working group.[1] This is followed by a discussion of the role of aviation and aeronautics and a listing of needs, opportunities, and their implications for aeronautics technology as determined by the working group that focused on the scenario throughout the workshop.

PUSHING THE ENVELOPE

U.S. economic competitiveness	☑ Strong	☐ Weak
Worldwide demand for aeronautics products and services	☑ High growth	☐ Low growth
Threats to global security and/or quality of life	☐ High	☑ Low
Global trend in government participation in society	☑ Low	☐ High

Summary

"Pushing the Envelope" is a scenario characterized by a continuously growing, strong global economy and an affluent and growing middle class. The United States also has a strong economy, with liberal trade policies and a laissez-faire attitude toward business and commerce. The worldwide demand for aeronautics and space-based products and services is strong and the free market economy is generating rising employment and prosperity throughout the world. Local and state governments are privatizing many of their services while opening up new markets

---

[1] The full texts of the scenarios developed by the NASA/TFG/SAIC core team are found in Appendix D.

for private investment. Technology has generated many new products and services while creating much shorter product life cycles. Trade barriers have crumbled as capital freely crosses international borders. In many cultures, market-based environmentalism is becoming a driving force. Quality of life is improving as consumers "dial in" on the Internet for more goods and services. The world is at peace and the role of the U.S. military is primarily one of deterrence and policing in such trouble spots as the Middle East. This role requires the military to focus on the protection of its personnel while using only enough force to control a situation without causing massive casualties. The military also relies on commercial assets to support many logistics and communications functions.

Aviation and Aeronautics in the "Pushing the Envelope" Scenario

An examination of the aviation marketplace reveals that the consumer is king. There is a worldwide demand for low-cost, efficient air service. Safety, performance, and customer service also are important. The increase in air travel has generated a growing demand for new aircraft products and services. Refurbished passenger and cargo aircraft are in high demand as manufacturers try to keep up with demands for new aircraft orders. Consumers, with their new-found wealth, are ordering more products through the Internet, which is creating a growing need for air cargo service. Business travel is at an all-time high, and transoceanic flights are generally overbooked. Emerging markets in China, India, South America, and Eastern Europe are generating large backlogs for new aeronautics products that will take manufacturers many years to fulfill. Strong, competitive forces dominate the marketplace as demand for services increases.

More and more business is being carried out face to face with customers and partners around the world because efficient capital markets and competitive market forces have generated many global alliances of convenience. Businesses also are sharing resources as they compete for growing market segments. Customer satisfaction is not just a slogan, but a requirement for business survival. The growing use of the Internet and advanced telecommunications has created a need for "just-in-time" products and services. Affordability, dependability, and reliability are the buzz words for a growing air cargo market. Prompt, efficient, and on-time service is a requirement for those companies that ship products by air.

Needs, Opportunities, and Their Implications

To satisfy this growing air travel market, several needs and opportunities were identified for aeronautics and are discussed below.

Worldwide Demand for Efficient, Low-Cost Air Passenger Service

The wide number of air travel choices available to consumers in this scenario would create a highly competitive marketplace and a demand for efficient, low-cost air passenger service. For air carriers, meeting this demand while remaining profitable would require low total operating costs, including low-cost systems and procedures related to aircraft maintenance, repair, and overhaul. Large-volume aircraft, reconfigurable aircraft, and supersonic transports also would be required to maximize the efficiency of air travel and meet the growing demands of both consumers and air carriers. Each of these aircraft types would be designed with both low production costs and low operating costs, and short cycle times for design, fabrication, and certification processes would be required for new aircraft products.

Customer-Tailored Air Products and Services

Another air carrier response to increased competition would be to offer customized services enroute that are based on the demands of the market segment. Examples include entertainment options, such as gambling for leisure travelers, or improved access to the office for business travelers. As a result, a wider variety of aircraft products, including reconfigurable aircraft, and closer involvement of customers in the design and manufacturing of new aircraft products, would be the norm.

Increased Investments in Existing Transportation Infrastructure

To handle the increased throughput of aircraft, people, and cargo that exists in the "Pushing the Envelope" scenario, investments in existing transportation infrastructure would need to be increased. New types of airports, including heliports, would be needed, and fast, efficient access to these airports would need to be found. Mass transit systems would need to be fully integrated with rotorcraft and tiltrotor operations by co-locating heliports with bus and rail stations and by coordinating passenger pick-up and drop-off schedules with aircraft arrivals and departures.

In highly developed nations, opportunities for new full-service airports would remain limited due to geographic and environmental constraints, as would increased operations at existing airports. This would create technological challenges to increase passenger and cargo capacity without greatly increasing the frequency of flights.

Improvements in the Air Traffic Management System

Improvements in ground-based air traffic management (ATM) integrated with on-board ATM would be required to handle the increased volume of air traffic with the efficiency and safety that would be demanded by the public. Automated air traffic control (ATC) using satellite-based communications, navigation, and surveillance systems with increased data-handling capabilities and improved weather sensing and modeling would be a must. Technology that allows aircraft to avoid adverse weather autonomously also would be required, and enhanced adverse weather landing capabilities would need to exist at most airports. In general, improved situational awareness systems would be required in the aircraft and on the ground.

High Security and Reduced Accident Rates

Although threats to security are generally low in this scenario, sophisticated sensing equipment and security measures would be needed to protect a large and growing number of air travelers. To reduce overall accident rates despite a large increase in the number of takeoffs and landings, aircraft would need to be designed, maintained, and operated in a manner that reduces human error. Advanced cognitive engineering techniques; automatic fault detection and correction systems; built-in, error-free system interfaces; and stringent safety procedures would be required.

Environmentally Friendly Air Transportation System

Although environmental protection is more market based in this scenario, it remains important to most consumers. Therefore, supersonic aircraft would require technology that would reduce sonic boom effects so as to make overland flights at supersonic speeds acceptable. In addition, all aircraft would be required to meet or exceed noise-level requirements beyond Federal Aviation Regulation (FAR) Part 36, Stage 3, to further reduce noise pollution.[2] Aircraft exhaust pollution emissions also would need to be greatly reduced.

Dependable, Reliable, Low-Cost Air Cargo Service

Twenty-four hour, on-time, adverse weather capabilities would be required for air cargo service providers so that shipments arrive on schedule. Modern facilities that

---

[2] FAR Part 36 (Amendment 36-21, 12/28/95), *Noise Standards: Aircraft Type and Airworthiness Certification* specifies the allowable levels of noise for transport and turbojet powered aircraft during takeoff and approach to airport runways. The exact decibel levels allowed vary by aircraft weight and number of engines.

enable fast and efficient cargo handling and sophisticated cargo-tracking capability through the use of common intermodal interfaces and advanced custom-tailored information technology would be needed. Safety and security requirements would be similar to those of passenger service.

Low-Cost, Reliable Access to Space

Competitive market forces would drive the commercial launch vehicle and satellite businesses toward low-cost service with increased reliability. Advanced air-augmented or air-breathing propulsion systems could be used on some launch vehicles if they proved to be cost effective. The U.S. military would continue to require dedicated space capabilities for global command, control, communications, and intelligence (C^3I), as well as navigation and surveillance. For both commercial and military space activities, the reduction of orbital debris would be a priority. This could create requirements for fully reusable launch vehicles that leave no components in orbit.[3] Launch vehicles that are robust enough to withstand debris impacts while in space also could be required.

Aerospace Research and Development Partnerships

Effective partnerships between academia, industry, and government would be required to maximize the efficient use of research and development (R&D) dollars. Increased use of peer review for the allocation of federal R&D funding would be a priority of this partnership, as would increased emphasis on technology verification and technology transfer.

Global Responsibilities of the U.S. Military

To support the global responsibilities that the U.S. military would have in this scenario, rapid reaction and deployment of military forces would be required. These deployments could require the substantial use of commercial assets for air transport and space-based communications and remote sensing. Adverse weather, modular, robotic, or autonomous smart weapons systems also would be needed for stand-off weapons delivery and defense. In general, emphasis would be placed on the reduction of casualties, the protection and survivability of high-value delivery systems, and the total cost and effectiveness of military operations. The use of

---

[3] Rocket bodies (such as upper stages) are estimated to be 17 percent of the total orbital debris population. However, measured by mass, rocket bodies are a larger portion of the total mass of the population and pose a threat as far as space debris is concerned because they are often left in orbits that intersect the orbits of functional spacecraft. Source: National Research Council. 1995. Orbital Space Debris—A Technical Assessment. Washington, D.C.: National Academy Press.

nonlethal weapons systems to achieve the required military objectives would be a highly desirable option.

GROUNDED

U.S. economic competitiveness	✓ Strong	☐ Weak
Worldwide demand for aeronautics products and services	☐ High growth	✓ Low growth
Threats to global security and/or quality of life	✓ High	☐ Low
Global trend in government participation in society	☐ Low	✓ High

Summary

The "Grounded" scenario is characterized by a world victimized by unprecedented, unpredictable, and anonymous large-scale violence and terrorist actions. A long period of rapid, global economic growth has served to widen the gap between the well-to-do, skilled, professional work force and unskilled and semiskilled workers, especially in developing nations. Angry, bitter, jealous people in socioeconomic groups that have been left behind are resorting in frustration to random acts of violence against large masses of people wherever they can be targeted. All forms of mass transit, large public gatherings, and dense urban environments are now avoided, if possible.

To keep the global economy healthy, the U.S. government has embarked on a massive program to significantly upgrade and expand the Internet system on a global basis. The new U.S. standard Global Information Network or "G-net" is now the worldwide standard.[4] With a massively increased bandwidth and a greater number of applications, the G-net has become the backbone of global commerce and an acceptable (and cost-competitive) alternative to all but the most essential travel. All aspects of society, including commerce, government, culture, and entertainment flow continuously and globally through the G-net. The global economy's vitality is sustained by the G-net and the rapid pace of technological advance, product innovation, and entrepreneurship that it enables. "Virtuality" has replaced "being there."

---

[4] In this scenario, the G-net represents an advanced version of the Internet.

Aviation and Aeronautics in the "Grounded" Scenario

Onerous and very expensive security measures, enacted to counter the threat of random violence and terrorism, have increased significantly the cost of most forms of public transportation, especially air travel. High cost and fear have combined to eliminate most discretionary air travel. For the small amount of personal air travel that is still conducted, unscheduled, point-to-point flights from more distributed, smaller airfields have replaced the large hub and spoke system of today. To more closely control security, most businesses operate their own aircraft fleets or charter small commercial jets in lieu of using commercial carriers. In many cases, families and individuals who must travel also feel more secure flying themselves or arranging for individual charters. As a result, general aviation also experiences some growth in this scenario.

Needs, Opportunities, and Their Implications

Specialized Cargo Aircraft

Even though the G-net enabled a global economy and strong U.S. economic competitiveness, the movement of goods over great distances would still be required, even though cargo shipments would occasionally be at risk from terrorism. This would create the need for a new, specialized, subsonic, large-payload cargo aircraft with a global range for time-critical cargo delivery to a fast domestic distribution system.

Low-Capacity, Long-Range Transports

Commercial and business travel that is not conducted using private fleets or chartered aircraft would be accomplished by air carriers using a new generation of lower-capacity transports that also operate within a more distributed, point-to-point air transportation system. The premium prices paid by those who still must travel would create a demand for premium services, including aircraft interior configurations customized to meet passenger requirements, and supersonic capability for long-range overwater routes.

General Aviation

The increase in general aviation (GA) activity would create a need for low-cost, reliable, user-friendly GA aircraft that operate much more like an automobile and have more adverse weather capabilities than today's aircraft. The G-net would play an important role in GA by enabling an extensive network of flight training and simulation capabilities to be delivered directly to the home. These capabilities would enhance both the safety and security of personnel by reducing exposure to actual flight hazards, including terrorism.

Military Operations Vehicle

Counterterrorist special operations and rapid response to regional flare-ups would emphasize the need for a stealthy, long-range, special operations forces vehicle that operates from unimproved fields with short take-off and landing capability and, in some cases, vertical capability.

Survivability and Security Technologies

For all classes of aircraft, the incorporation of a wide range of survivability features would be required to maximize customer safety. In many cases, these systems would be derived from military activities and research programs. Aggressive technology development activities that benefit both military and commercial requirements would be needed so that many of these features would be affordable for general application. These features include a full range of threat detection and mitigation systems, including bomb detectors and missile launch detectors, missile countermeasures such as infrared signature reduction, and electromagnetic interference and laser aircraft hardening. Threat avoidance operations that use high-lift or vertical-lift capability to minimize exposure to ground-based threats (such as shoulder-mounted missiles and small arms fire), and survivable aircraft features (such as reconfigurable controls, adaptable control surfaces, and survivable structures) also would be needed.

Ground security measures would be extensive and would include expanded airport security perimeters (i.e., gated entrances to airport property), as well as improved screening of personnel, passengers, cargo, and baggage. The real-time access to worldwide databases on the G-net would further enable a rapid, highly reliable system to identify and profile all passengers.

Air Traffic Management

Point-to-point operations that use lower-capacity commercial transports, business jets, and GA aircraft would need to be supported by a sophisticated distributed airspace system with automated decision making for ATM. Technology to provide accurate, user-friendly situation awareness and decision aids directly to the cockpit by way of secure and reliable data links would be needed.

Access to Space

To establish and maintain the network of communications and navigation satellites inherent in this scenario, a low-cost, on-demand, low-Earth orbit (LEO) launch capability is paramount. A new generation of extremely small, lower-weight satellites would allow the development of a reduced-cost, unmanned orbital insertion vehicle.

Enhanced Modeling and Simulation

The computational and networking capabilities that exist in the "virtual" environment of this scenario would enable an entire spectrum of modeling and simulation capabilities that could transform many industries, including aeronautics. Product development lead times and manufacturing costs would be slashed, and an entire new standard of reliability would be established, provided that technology is developed to

- improve the modeling of aerodynamics, propulsion, and structures using fundamental physical laws rather than empirical data
- improve process modeling and human-machine interface modeling
- enable virtual manufacturing, assembly, and even testing

REGIONAL TENSIONS

U.S. economic competitiveness	☐ Strong	☑ Weak
Worldwide demand for aeronautics products and services	☑ High growth	☐ Low growth
Threats to global security and/or quality of life	☑ High	☐ Low
Global trend in government participation in society	☐ Low	☑ High

Summary

The future world of "Regional Tensions" is characterized by weak U.S. economic competitiveness, a high growth rate of worldwide demand for aeronautics products and services, a high level of threats to security and quality of life, and a high global trend in government participation in society.

The world has organized itself into regional trading blocs, ending a period of harmonious globalization. Japan, Russia, and India have formed an alliance so as to compete with China. An Anglo-American bloc has emerged including Australia, New Zealand, Taiwan, the United States, and the United Kingdom. The remainder of the European Union takes no sides and attempts to sell its goods to any nation.

Rapid change and regionalization have made the world uncertain and uneasy, although it is still at peace and has nominally open trade practices. The United States has lost many of its markets and manufacturing capabilities within the competing regions. The United States is in an ambiguous position, with the need to prepare for potential hostilities because of regional tensions while also recognizing the need to strengthen its economy by increasing trade within its region and restoring trade with other regions. The nation's strategic goals in this scenario, as determined by the world team, are to protect home regions, ensure the supply of critical natural resources and energy, prevent world war, contain expansion by potentially hostile nations, recover interregional markets, and unify U.S. society.

Aviation and Aeronautics in the "Regional Tensions" Scenario

As a result of the goals stated above, the United States has turned its attention inward and allowed the government more control so as to mobilize the nation for action. The government is investing its aeronautics R&D resources almost entirely in military projects, although it insists that these resources have an inherent dual civilian use. The government also is expanding aeronautical education programs so as to use domestic talent to replace the skills lost to overseas competitors.

Several features of this scenario affect commercial aviation. The International Civil Aviation Organization (ICAO) and the World Trade Organization (WTO) are no longer functional, having removed their organizing influence in their respective domains affecting aeronautics. Oil is in sufficient supply, but at a high enough cost to encourage fuel conservation. Nevertheless, the tempo of international air travel is high. Within the United States, little capital is invested in ATM and related infrastructure. However, the dispersion of industrial facilities for both military and economic reasons has increased the requirement for flexible air traffic routing to and from minimal-sized, distributed landing fields.

Environmental issues have a reduced priority in society and therefore have little impact on aviation. However, the global spread of infectious diseases that are attributed to the high tempo of international air travel is an ongoing concern.

Needs, Opportunities, and Their Implications

The global situation in the "Regional Tensions" scenario, as reflected in the U.S. strategic goals, would result in the following compilation of needs that would have consequences to aeronautics:

- regional industrial capabilities to produce a wide variety of defense and commercial vehicles at reduced cost but without low-cost labor
- a skilled domestic labor force, including engineers, scientists, and shop workers trained by an education system that is focused on achieving broad-based participation from all segments of a diverse society
- a rapid, reliable, and flexible distribution capability for manufactured products
- an ability to project military force
- alternative domestic energy sources
- secure, reliable communications and information systems
- military and industrial intelligence gathering capability
- air transportation systems that enable operations in both existing, sophisticated infrastructures and new, but austere infrastructures
- appropriate defensive military capability
- prevention of the spread of infectious diseases by global air travel

Eight system level aeronautics technologies have been identified that would meet these needs. They are summarized below.

Unstaffed or Autonomous Air Traffic Control

To produce aircraft that can operate efficiently in domestic service without investment in new ATC infrastructure and would also be attractive to customers in overseas markets with austere infrastructures, an autonomous, self-contained airborne ATC system would be desired. It would employ on-board information systems that use cooperative aircraft position information that is broadcast by satellite. These systems would provide navigation, collision avoidance, terminal landing, and adverse weather avoidance capability.

Example technology requirements are increased bandwidth capacity for reliable communications and data management, reduced pilot workload, improved computer algorithms for collision avoidance, and integration within the existing ATC system.

Virtual, Smart, Universal Aircraft "Factory"

To make the best use of scarce resources and compete strongly for foreign market share, a new type of aircraft factory would be needed that builds on the knowledge and information technology strengths available in the United States. This factory would be "virtual" in the sense that the contributing units would operate as a cohesive whole even though separated physically for security and economic reasons. The products manufactured by the factory would include many types and sizes of aircraft, providing variations of basic and reconfigurable systems. Product attributes would be optimized by means of total system optimization design tools.

The principal technology requirements would be integrated product design, manufacturing, and certification capabilities that enable new aircraft to be designed, developed, manufactured, and sold within one year[5]; integrated learning systems for equipment, products, and manufacturing processes; and rule-based "expert" systems that support innovation and universal tooling.

Short- to Medium-Range, Infrastructure-Independent Transports

To penetrate emerging markets within our region and undeveloped portions of other regions, a vehicle needing minimal-sized landing areas, with autonomous ATC and austere operating capabilities, would be extremely desirable. This aircraft also would serve as a multirole (e.g., special forces) military transport and could be used for the short-haul, dense domestic market to support dispersed manufacturing facilities with minimal impact on the existing infrastructure. Vertical/short take-off and landing technologies would be required for this aircraft, as well as integrated propulsion and airframe approaches to noise reduction; low specific fuel consumption, high thrust-to-weight ratio engines; autonomous ATC; and advanced manufacturing techniques.

Extremely Long-Range, Large, Supersonic Transports

To project military force and distribute high-value industrial products efficiently, a global range, supersonic transport would be required. The aircraft's high velocity would increase its productivity by increasing the amount of cargo or personnel that could be delivered within a given period of time. When used to carry passengers, the large seating capacity would reduce the burden on the ATC system and airport infrastructure. Technology to reduce the aircraft's sonic boom while traveling over land would be required, as well as non-afterburning supercruise engines and low

---

[5] The steering committee recognizes that this is a challenging goal. However, it could conceivably be achievable in 15 to 25 years if the required design and manufacturing technologies are successfully developed.

supersonic drag aerodynamic configurations, which could possibly be achieved by a modification of the oncoming airflow with beamed energy.

Skill-Targeted Education System

To capitalize on the existing domestic workforce, the number of college graduates with degrees related to aeronautics would need to be increased dramatically. Universities would have to provide an initial core education (to allow flexibility in later years), followed by specific training that uses industry as the campus. This would be both supported and facilitated by government funding.

Uninhabited Air Vehicles

Uninhabited air vehicles (UAVs) would be employed for a number of military and civilian activities, including communications, surveillance, intelligence gathering, local weather observation, aerial combat, and even cargo transport. UAVs would be chosen for these missions because they offer the best combination of long-term endurance, long range, low cost, and low observability, while removing the pilot and crew from exposure to hostile environments. The key technology requirements are improved understanding of human-machine interactions; reliable and secure communications links; flexible and modular payload configurations; low fuel consumption engines for high-altitude operation; and inexpensive, flexible manufacturing techniques.

Controlled Observability, Low-Noise Technologies

Controlled observability and low-noise technologies are key areas of technology in their own right, and together are usually referred to as stealth technology. In this scenario, the military would apply the knowledge gained from years of advanced stealth technology R&D to the design and production of weapons systems that can locate, target, and destroy enemy stealth aircraft. On the commercial side, some stealth technology would be desirable to minimize the threats to passengers posed by terrorist missile attacks. Stealth technology also has a secondary benefit of reducing the impact of aircraft noise and emissions on airport operations.

Low-Cost, On-Demand, Nonhuman Access to Space

To supply affordable satellite-based communications, navigation, surveillance, and weather forecasting capabilities to both the military and the civilian sectors, inexpensive payloads and launchers would be required. For the military, on-demand launch capability would be needed to overwhelm the ability of an enemy to destroy our satellites. For the civilian sector, industrial espionage may also require on-demand launch capabilities. Low-cost, launch-on-demand vehicles may utilize air-breathing propulsion while traveling through the oxygen-rich portion of the atmosphere. The key technology requirements include novel air-breathing propulsion devices such as gun-launched ramjets, scramjets, and ram accelerators; inexpensive expendable components; and technology to provide aircraft-like operations.

TRADING PLACES

U.S. economic competitiveness	☐ Strong	✓ Weak
Worldwide demand for aeronautics products and services	✓ High growth	☐ Low growth
Threats to global security and/or quality of life	☐ High	✓ Low
Global trend in government participation in society	✓ Low	☐ High

Summary

In the "Trading Places" scenario, Asia is the dominant economic power in a secure and stable middle-class world. The greatest capital resources and buying power reside in China, along with the greatest potential for market growth. The United States is relatively weak in its economic power. Threats to security or aggressive actions are not significant elements of this world. Although not active, China has military power that exceeds other nations' power. Governments of the world play a value-added role in society that is "allowed and controlled" by consumers.

In this world of global business and activity, China has become a significant marketplace and an economic power. This rapidly developed and still growing nation that is spread over a large area has bypassed the development of a surface transportation and communications infrastructure and now has a population that is predominantly middle class.

Consumers in this scenario are generally middle class and seek a quality of life that causes the population to be as geographically distributed as possible. They also respect and pursue environmental cleanliness. In addition, middle-class consumers experience a scarcity of leisure time. Nevertheless, Asians prefer to travel to the United States for vacation.

Business in this scenario is competitive globally, and international teams and partnerships are the norm. Asia has become the manufacturing center for products required to sustain both its own population and its economic strength through export.

Aviation and Aeronautics in the "Trading Places" Scenario

Because China has bypassed the development of an extensive ground-based infrastructure, the transportation needs of its industries and middle-class consumers are mostly met by aviation. Regions rely almost entirely on air transportation systems that are tailored to serve their regional transportation needs. Industries provide "commute to work" services to geographically dispersed employees with privately owned and operated air transportation systems. Industries also use their transportation systems for the movement of raw and finished goods. Leisure travel by air includes point-to-point types of flights (i.e., Beijing to Jackson Hole, Wyoming) that must be comfortable and fun. The travel itself actually is considered part of the vacation's leisure time.

Both the teaming nature of business and the need to maintain an understanding of foreign markets creates intense demand for high-speed travel and communications that are used extensively.

Needs, Opportunities, and Their Implications

This scenario provides an opportunity to start with a clean sheet of paper and rethink the ways the overall air transportation system can be optimized for specialized markets. The needs and opportunities that result are discussed below.

Privately or Regionally Owned and Operated Air Transportation Systems

This somewhat new application of aeronautics would require air transportation systems that meet criteria different from those of the system that currently exists today.[6] Privately or regionally owned air transportation systems would need to be tailored for both customers and for multiple uses within each customer's system. To support this need for flexibility, the system would need to be quickly reconfigured to accommodate multiple payloads, performance characteristics, and ground-based infrastructures (austere to sophisticated). Consequently, the design,

---

[6] In this scenario, air transportation system is defined as all the elements required for air transportation, including aircraft (subsonic and supersonic jets, rotorcraft, general aviation aircraft, etc.), landing provisions (airports, landing pads, etc.), air traffic management, communications, maintenance and logistics, and flight crews.

integration, and allocation of all system elements to be flexible, cost effective, and tailored for specific customer needs would require the "clean sheet of paper" approach. In addition, safety would need to be "designed" into the air transportation system rather than "inspected in" after the fact. This would allow the system to operate as safely or even more safely than today's air transportation system with less reliance on periodic inspections. Specifically, the system would contain a self-diagnosing capability that would terminate operations when unsafe conditions exist. Reliance on human skills and expertise would have to be minimized to ensure safe operation by regional groups and private individuals who were not career aviation experts.

Technical capabilities that support the need for dependable and safe air transportation systems would include smart systems and reconfigurable systems; on-site repairability that uses standard equipment; required skills training and repair support provided by communications capabilities that are similar to Internet[7]; immunity to most weather and runway conditions; and simple, affordable runway operations (e.g., short runways, inexpensive runways, or runways with multiple uses). Affordable and reconfigurable aircraft would require modular designs with snap-together, load-bearing disconnect–attach mechanisms and wireless, open-architecture avionics systems. Reconfiguration and safety requirements would be satisfied with smart systems, structures, and materials that detect breakdowns or problems and respond with autonomous adjustments that ensure continued safe operation or halt operations if safety cannot be assured.

Long-Range, Large-Payload Aircraft

Comfortable, fun, and affordable travel from locations throughout China directly to locations in the middle of the United States would create a need for very large, efficient aircraft. Providing desired and valuable leisure time for passengers while enroute would place less emphasis on aircraft speed and more emphasis on quiet, humidified, spacious aircraft interiors that would allow passenger activity. Entertainment on long flights would require extensive communications capabilities for the aircraft as well. The middle-class quality-of-life standards expected by Asian travelers in this scenario would create demands for safety, reliability, and environmental friendliness that equal or exceed today's aircraft performance in these areas.

The successful development of this aircraft would require significant improvements in airframe configurations, materials, systems architectures and technologies, and

---

[7] Such a communications help line would provide readily available, accessible computer network-based interactive training courses for pilots, maintenance personnel, and people in related operations. Examples of specific capabilities include flight simulation, student certification, and on-line maintenance troubleshooting.

approaches to communications technology. New structural materials that provide multiple features would help to achieve the required performance. This could be a material that provides structural support, and also acts as thermal and acoustic insulation for cabin areas, creates a flame barrier for the protection of passengers, and is immune to condensation and subsequent corrosion caused by increased cabin humidity.

An aircraft with similar levels of performance advancement in payload size, range, safety, reliability, and environmental friendliness also would be required to meet the cargo transport needs of this scenario's global marketplace.

High-Speed Business Travel to Any Location in the World

To meet the travel demands of business people who require personal interaction with partners, customers, and clients around the globe, high-speed air travel capability would be needed in this scenario. The rapid movement of people or cargo is also important for resolving unplanned or emergency situations. Supersonic aircraft would not be optimized for leisure travel, but would need to be designed to minimize environmental impacts such as noise and emissions over all land masses.

ENVIRONMENTALLY CHALLENGED

U.S. economic competitiveness	☐	Strong	☑	Weak
Worldwide demand for aeronautics products and services	☐	High growth	☑	Low growth
Threats to global security and/or quality of life	☑	High	☐	Low
Global trend in government participation in society	☐	Low	☑	High

Summary

The "Environmentally Challenged" world is one in which strict limitations are placed on worldwide carbon dioxide (CO_2) emissions because of conclusive evidence showing it harms the planet. Limitations and concerns are strongest in the developed nations; Europe is first to recognize the CO_2 problem. The United States, with high per capita energy consumption and emissions, is at a distinct disadvantage. Although the developing nations accept that CO_2 emissions are a problem, they place most of the blame on industrialized nations and are reluctant to limit emissions if it hampers their own economic development and growth. Therefore, some developing nations seek to circumvent or evade emissions limitations. There is a system of caps on CO_2 emissions for each nation that is enforced by international organizations. In the United States, there is also a system of tradable pollution rights and very high taxes on carbon-based fuels.

The United States is a weak competitor in this scenario because of its high reliance on energy-intensive industries. The nation's large land area requires heavy use of transportation systems, which makes the United States a more intensive user of fossil fuels than most other nations. Because of high fuel prices, there is low growth in the demand for aerospace products and services. The developed world, including Europe, Japan, and the United States, imposes trade and other sanctions on lesser-developed, noncompliant nations in an attempt to force compliance with the CO_2 emission limits. This strategy raises tensions between the "have" and the "have-not" nations and has a destabilizing effect on global security.

The strong need to stop damaging the environment has captured public attention in the developed world. There are sporadic terrorist threats and attacks on intensive users of carbon fuels, such as the air transportation industry. The weak U.S. economy exacerbates domestic tensions with some of the nation's population looking to technology as a potential solution, whereas others view technology as a root cause of the problem.

The consumer prices for all modes of transportation are high because of high taxes on carbon fuels. In an attempt to reduce fuel consumption, airlines are shifting to less frequent service with larger aircraft and higher load factors. Aircraft also are now operated at optimized speeds (usually lower than current cruise speeds) and fly optimized flight paths so as to minimize fuel consumption. Firms also seek to reduce the energy input of producing goods and services by revamping production processes and by reducing the use of transportation for people and goods. As a result, both business and pleasure travel are reduced significantly.

Aviation and Aeronautics in the "Environmentally Challenged" Scenario

The aerospace industry, facing a much lower demand for its products, aggressively seeks new markets. Distinct markets evolve for retrofit technology to reduce the fuel consumption and the emissions of existing commercial aircraft fleets without purchasing entirely new aircraft at greater expense. The U.S. aerospace industry is lagging compared with Europe's aerospace industry because it was late in increasing energy efficiency and reducing emissions in both its production processes and its product goods. There is a market demand for improved ATM to reduce fuel consumption as well as for sensors to detect pollution. U.S. firms are trying to exploit foreign markets where economic growth is stronger, but manufacturers and suppliers are forced increasingly to pursue alliances with foreign companies so as to gain entry to these foreign markets. In addition, U.S. companies often bring only technology to the bargaining table because of the scarcity and cost of investment capital in the United States.

The U.S. market for military aeronautics products also is characterized by low growth in this scenario. Nevertheless, there is a heightened demand for monitoring and counterterrorism systems, including satellites and UAVs. There also is a growing demand for microvehicles (tens of kilograms) and systems that operate with extremely low fuel consumption. The U.S. military uses these systems to monitor other nations for reasons that include maintaining U.S. security, determining environmental compliance, and enhancing U.S. economic competitiveness.

Needs, Opportunities, and Their Implications

In this scenario, the air transportation industry and the aerospace industry are particularly challenged because of their high-energy requirements. High fuel costs and financial incentives to reduce emissions would create the following needs:

- aircraft with reduced or zero CO_2 emissions
- the ability to place remote sensing systems for CO_2 monitoring and security into LEO and into the atmosphere on demand and at low cost
- ATM improvements to reduce fuel consumption

In addition, the high level of tension would create threats to civil aviation that would require improvements to the security of commercial aircraft. The technological implications and R&D needed to satisfy these needs are discussed below.

Aircraft with Reduced or Zero CO_2 Emissions

There are two research alternatives that could lead to the development of aircraft with either reduced or zero CO_2 emissions. The zero emissions objective would require the development of engines that burn nonfossil fuel, such as hydrogen-powered engines. It also would require that the energy used to produce hydrogen is not itself a source of CO_2 emissions. Nuclear energy or the hydroelectric generation of electric power are two possibilities. In addition to the hydrogen-fueled engine, the following capabilities and technologies also would be required to enable the operation of hydrogen-powered aircraft:

- low-weight cryogenic materials for hydrogen fuel tanks
- efficient means to produce slush hydrogen on or near airports
- safe means of handling hydrogen, including leak detection and mitigation

- efficient aircraft configurations to accommodate the size and location of hydrogen fuel tanks

Emissions reduction could be achieved by the development of technology to reduce significantly the fuel consumption of conventionally fueled aircraft. This alternative would have the advantage of being applicable to not only new production aircraft, but also would offer the potential for retrofit to the existing commercial aircraft fleet. Aerodynamic improvements to increase lift and reduce drag, such as microelectro-mechanical systems for flow control, would be one area of research to pursue. New, lighter "tailored" (designed at the molecular level) and "smart" (able to sense their own conditions) materials to reduce aircraft weight without compromising structural integrity would be another area of investigation. More fuel-efficient engines, which would require the development of higher-temperature materials, would be a third area. In addition, there would be a need to develop the means to remove CO_2 from conventional jet engine emissions. These could be biologically based systems or chemical processes that break down CO_2 into nonharmful components.[8]

Low-Cost, On-Demand, Remote Sensing Capabilities

In the "Environmentally Challenged" scenario, there would be a need to monitor compliance with emissions regulations and to counter selected security threats to the United States. These objectives would require the development of improved remote sensing systems that would operate from platforms within the Earth's atmosphere and in orbit. UAVs, which could carry sensor payloads within the Earth's atmosphere, would require improvements in data links, autonomous decision making, and related control technologies. Many of these sensors would be small and lightweight (on the order of tens of pounds), but would have the same functionality as larger sensors used today. Placing these sensors into Earth orbit on demand and at low cost would probably require technology developments for both payloads and launch vehicles.

Improved Air Traffic Management to Reduce Fuel Consumption

This scenario, as mentioned previously, includes high fuel costs and limits on emissions. The development of improvements to ATM technology to enable the most fuel-efficient routing for aircraft could provide large benefits for commercial air carriers. To achieve this goal, technology would be required that improves the accuracy of sensors used for navigation and surveillance, improves the accuracy of

---

[8] The steering committee believes that developing this capability may prove to be impractical. However, it is certainly an area of research that would need to be pursued in this scenario.

weather monitoring and weather information dissemination to aircraft, and improves overall security and integrity of the ATM system.

Improved Security for Commercial Aircraft

Technologies to improve the security of commercial aircraft operations would be required in this scenario. Requirements would include the adaptation of low observable and missile countermeasures technology to commercial aircraft without impacting negatively overall aircraft safety, cost, or performance. The structures and materials used in commercial aircraft also could be made more resistant to on-board bomb blasts or munitions from external sources. Finally, contraband detection systems could be integrated into the aircraft as a means of enhancing passenger safety and security.

SUMMARY OF NEEDS AND OPPORTUNITIES

Each need and opportunity mentioned in one or more of the scenarios presented above is included in Table 2-1. In many cases, a need or opportunity that was found to be important in one scenario also was important in another. However, the exact characteristics of each need or opportunity varied from one scenario to the other. Table 2-1 attempts to illustrate this by using generic language for the needs and opportunities listed in the far left-hand column, while using language in the remaining five columns that corresponds to the context for the need or opportunity within each scenario. A blank space in the table indicates that the need or opportunity in the far left-hand column was not considered in the given scenario.

TABLE 2-1 Summary of Needs and Opportunities for Each Scenario

Needs and Opportunities	"Pushing the Envelope"	"Grounded"
Air Traffic Management (ATM)	Worldwide, high volume of air traffic	Distributed airspace system
Access to Space	Low-cost commercial launch, dedicated military launch	Low-cost to low-Earth orbit (LEO)
Airport Infrastructure	Expanded in third world, existing infrastructure constrained in developed world	Distributed non-hub system
Safety and Survivability	Reduced accident rate for increased air traffic	Survivable structures, adaptive controls
Manufacturing	Short cycle times, low production costs	Enhanced modeling and simulation
Air Cargo	Large-volume, reconfigurable aircraft. Low-cost, reliable cargo tracking and handling	Specialized, subsonic, large-payload aircraft
Uninhabited Air Vehicles (UAVs)	Deliver smart, stand-off weapons	
Environment	Low noise and emissions	
Short- to Medium-Range Aircraft		
Security Systems	Increased security	High-threat security systems
Supersonic Aircraft	Reduced noise and pollution	Long-range, low capacity aircraft
Stealth Technology		Military special operations aircraft
Subsonic Aircraft	Low-cost, large volume, reconfigurable aircraft	Low capacity for passengers, large payload for cargo
Vertical/Short Takeoff and Landing (VSTOL) Aircraft		Military special operations, long-range, stealth aircraft
General Aviation (GA)		Increased activity; low-cost, reliable, user-friendly aircraft
Tailored and Smart Materials		
Microelectro-mechanical Systems		
Training and Education		Distributed flight training

"Regional Tensions"	"Trading Places"	"Environmentally Challenged"
Autonomous, self-contained ATM on board aircraft	Integrated part of flexible, cost-effective, customer-tailored air transportation systems	Optimized to reduce fuel use
Low cost, nonhuman, launch on demand		Low-cost, on-demand launch capability for remote sensing payloads
Minimal-sized, distributed landing fields	Austere to sophisticated, distributed and tailored	Must accommodate hydrogen-powered aircraft
Reduced pilot workload, improved collision avoidance; UAVs to reduce human exposure to hostile environments	Safety designed into systems, significantly improved	Bomb and blast resistant aircraft materials
Virtual, smart, universal "factory"	R&D to production modeling and simulation capability	Energy efficiency is paramount
	Reconfigurable aircraft, supersonic aircraft	
Military and civilian applications		Used for remote sensing applications
	Environmentally optimized aircraft (low emissions and noise)	Reduced or zero emissions aircraft
Infrastructure independent, short- to medium-range, VSTOL military special operations	Part of customer-tailored air transportation system	
		Security technology for aircraft
Extremely long-range, large, reduced sonic boom aircraft	Long range for business travel and cargo hauling, low noise and emissions over populated areas	
Controlled observability and low noise for military and commercial aircraft		
	Long-range, large payload, and optimized for entertainment	Designed and operated for fuel efficiency and reduced or zero emissions
Infrastructure independent, short- to medium-range, military special operations	To accommodate austere landing conditions	
	Part of tailored air transportation system	
	Multifunction materials	Airframe (lightweight) and engine (lightweight, high temperature) materials for fuel efficiency
		Reduced size and weight of aircraft and sensors
Skill-targeted aeronautics training and education	Simplified training, accomplished by Internet-like communications network	

3

Needs and Opportunities, Technology Implications, and the Future Role of NASA

FUTURE NEEDS AND OPPORTUNITIES

Although there were differences among the five scenarios, future needs and opportunities for aeronautics and the air transportation system, in their broadest context, were not found to be revolutionary. People, cargo, and weapons will still need to be transported by air throughout the world over various distances and at various speeds. The systems, vehicles, and subcomponents required to meet those needs, however, may demand significant technological breakthroughs and therefore are in themselves revolutionary. Examples include high-speed civil transports using alternative fuels or very large subsonic aircraft using more lift-efficient wing forms that are still small enough in span to allow for operations at existing airports with existing runway structures.

There are many areas of aeronautics technology that, although not revolutionary, met critical needs in one or more of the five scenarios or provided substantial opportunities for the air transportation sector. Therefore, these technologies must be included in the future aeronautics R&D portfolio of the United States. Some examples include lighter, stronger, and safer materials; engines with reduced fuel consumption and environmental impact; and improved air traffic management. Furthermore, there is a need for more emphasis on technology validation and verification to allow more rapid product certification.

Revolutionary changes in operating principles also may be required to implement many of the aeronautics technologies and technical systems required to meet future needs and enable future opportunities. For example, the widespread implementation of autonomous ATM will require substantial changes in philosophy for many users and operators of the current ATM system. Acceptable global standards for technology and procedures that could accommodate different political and social cultures also would be required. In many ways, these challenges are greater than the technological ones.

Robust, Significant, and Noteworthy Needs and Opportunities

The workshop participants utilized the iterative round-robin process described in Chapter 1 to distinguish between future needs and opportunities that were "robust," "significant," or "noteworthy." Table 3-1 illustrates this distinction for each type of need or opportunity identified by each world team and discussed in Chapter 2. Robust needs and opportunities were those cross-cutting items that fit within the environment of every future scenario. Significant needs and opportunities were critically important to three or four of the scenarios. Noteworthy needs and opportunities were items that were novel and, although important, applied to only one or two scenarios.

TABLE 3-1 Robust, Significant, and Noteworthy Needs and Opportunities

ROBUST Common to all scenarios	SIGNIFICANT Less common but vital to some scenarios	NOTEWORTHY Specialized and unique
Air Traffic Management satellite-based, autonomous, tailored	Access to Space small payloads, low cost, on demand	Short-to-Medium Range Aircraft VSTOL, commuter, infrastructure independent, military special operations
Airport Infrastructure constrained, austere, tailored	Supersonic Aircraft long range, large, and low capacity	Stealth Aircraft evade terrorist threats, quiet over populated areas
Safety/Survivability significant accident reduction, survive natural and man-made threats	Subsonic Aircraft large, small, long and short range Air Cargo large, low-cost, specialized and reconfigurable aircraft	General Aviation increased activity, part of a customer-tailored air transportation system
Manufacturing agile, virtual, validation, certification	Uninhabited Air Vehicles weapons, surveillance, intelligence	Tailored and Smart Materials reduced fuel consumption and enhanced safety
	Environment noise, emissions, hydrogen fuels	Microelectro Mechanical Systems reduced fuel consumption and vehicle size
	Security Systems airport, aircraft, terrorist threat	Sonic Boom Mitigation enable supersonic flight over populated areas
	Vertical/Short Takeoff and Landing (VSTOL) Aircraft short, medium, and long range, stealth, infrastructure independent, military special operations	
	Skilled Training and Education distributed and tailored training	

Findings

Challenging Needs and Opportunities

Scenarios that reflected a favorable future for air transportation and aeronautics, such as "Pushing the Envelope," place challenging demands on technology. For example, an air transportation system that is characterized by higher volumes of traffic, tightly constrained operating environments, austere and sophisticated infrastructures, improved efficiency and flexibility, and lower accident rates will require challenging evolutionary advances in ATM technology. Aircraft that are more capable or combinations of aircraft types with greatly increased cargo and passenger capacity that can still operate out of existing airports also will require significant technological advances.

Specialized Needs and Opportunities

Many of the scenarios placed a significant amount of stress on the air transportation system because of the severe operating environments characterized by terrorism, environmental degradation, and regionalization. These conditions create specialized needs and opportunities. For example, reduced business and personal travel, combined with a need for greater long-haul cargo capacity, will require a more decentralized point-to-point air transportation system rather than a hub-and-spoke system. Stronger, more regionalized economies in places with no transportation infrastructures, such as in some developing nations, create an opportunity to develop and export "total air transportation" systems that include aircraft, ATM, maintenance, logistics, and all related infrastructure in one complete package.

Military Needs

Regional conflicts in most of the scenarios created military needs that included rapid force projection, surveillance, intelligence gathering, and information processing capability. The likelihood of a major global confrontation was low in all the scenarios, but the need for deterrence was still required to provide defense against a rising or resurgent superpower or a rogue nation that has acquired a weapon of mass destruction.

Access to Space

The need for access to space was driven in four of the five scenarios by a desire for low-cost, launch-on-demand vehicles that could carry small satellites (generally less than 500 kilograms) into Earth orbit for a variety of applications including communications, navigation, surveillance, and intelligence gathering; weather observation to support aviation and provide severe storm warnings and environmental monitoring; and emerging commercial applications in market areas such as agriculture, resource exploration, and land management. Low cost was the overriding requirement for commercial applications, whereas military applications required assured, rapid, and frequent launch-on-demand capabilities. Unmanned launch vehicles were overwhelmingly preferred.

Manufacturing Needs and Opportunities

In many of the scenarios, it appeared that the aerospace manufacturing infrastructure would need a technology stimulus to maintain the sector's long-term economic competitiveness. An industrywide focus on short-term needs left only the government to support high-risk manufacturing technology that would permit flexible and agile responses to rapidly changing markets and would enable modular production and assembly capabilities. Manufacturing processes that relied on significantly improved modeling and simulation capability were seen as an opportunity to lower the cost and shorten the production cycle for complex products and services. Similarly, many of the scenarios pointed to a need for continued government stimulus of education and training to support a more modeling- and simulation-oriented manufacturing base and to guard against an offshore brain drain and a loss of skilled technicians.

SYSTEM LEVEL TECHNOLOGY IMPLICATIONS

The steering committee's synthesis of the needs and opportunities discussed in the previous section, and their implications for broad areas of technology development, is provided below. The system level technologies identified under each of the six headings do not appear in order of priority and have not been comprehensively analyzed to determine their relative scientific merit or technical feasibility. They simply represent the principal items discussed at the workshop based on an analysis of the five scenarios and the iterative round-robin process. Further analysis will be needed to justify the spending of scarce R&D funds on many of these areas of technology.

New Aircraft

Rising economies in Asian nations and elsewhere in the developing world are producing new demands, often in spite of the lack of infrastructure such as airports, air traffic control systems, weather and other information services, and logistics and maintenance facilities. These new markets are likely to demand new types of aircraft. In particular, the steering committee identified a need for technology development focused on the following:

- short-range cargo and passenger aircraft, with attention to short takeoff and landing and other capabilities for operating within an austere infrastructure
- long-range, high-capacity supersonic aircraft
- modular and reconfigureable aircraft to accommodate various mixes of passengers and cargo or to accommodate both military and civilian functions
- aircraft with on-board air traffic control capability that can operate relatively autonomously with minimal ground support
- aircraft with on-board repair and maintenance capabilities that offer redundancy, self-inspection, and repair of electronics and other systems
- aircraft that utilize smart systems, structures, and materials that can detect damage in critical airframe components and respond with autonomous adjustments to ensure safety

System Integration in Aircraft Design, Manufacturing, and Operations

To maintain its competitive position, the United States will need to foster:

- the integration of mathematical models (physical, economic, and human), virtual reality,[1] and other methods for visualizing and evaluating designs, including planning tools for agile and flexible manufacturing
- the combination and integration of avionics and other information systems within aircraft and within the air traffic control, aviation weather, and maintenance and repair facilities to ensure reliable communications and efficiency of operation

Safety and Security of Passengers and Crews

Increasing demand for operational safety, combined with public awareness of the threat of terrorism and increasing air traffic at major airports requires:

- improved weather observation, forecasting models, and the capability for real-time information dissemination to end users such as pilots and air traffic controllers

---

[1] Virtual reality is a computer-based technology that allows the user to interact with data that give the appearance of a three-dimensional environment or world. The user can "enter" and "navigate" the three-dimensional world portrayed as graphic images and interact with objects in that world as if "inside" that world.

- on-board, user-friendly decision aids to improve situational awareness and to mitigate human error[2]
- improved aircraft system reliability through fault tolerance and artificial intelligence that would adapt to events such as the loss of critical control surfaces
- improved systems within aircraft to detect explosives and contraband
- improved aircraft survivability to bombs, missiles, small armaments, laser beams, electromagnetic impulses, and radio frequency interference, and to weather phenomena such as lightning strikes or wind shear

Improved Operating Efficiency and Cost Effectiveness

The United States can improve the cost effectiveness and operating efficiency of both civil and military aviation by pioneering the following capabilities:

- Increased automation of aircraft control and ATM, supervised by human pilots and ground controllers through a much improved global positioning system (GPS) and other sensors and wideband, highly reliable communication. This includes autonomous concepts for enroute as well as terminal operations, both in the air and on taxiways.
- UAVs to capitalize on the benefits of performance, cost, and crew safety that are made possible by removing humans from the aircraft, which would eliminate their chance of being injured in a crash or captured by adversaries. Initially these vehicles will be used for surveillance and weather observation, but eventually they might be used for aviation applications such as aerial combat and cargo transport.
- Tailored materials (designed at the molecular level) and smart materials (able to sense their own conditions) with predictable properties that reduce aircraft weight and substitute for current components. This would include airframe materials that combine functions of load bearing, thermal insulation, and vapor impermeability, as well as engine materials to withstand higher temperatures.
- High-efficiency subsonic propulsion systems that provide improved fuel consumption.
- Miniaturization of electronics, sensors, and other nonstructural mechanical components to reduce weight and enable new aircraft and propulsion system designs.

---

[2] Situational awareness is a pilot's or crew's awareness of flight conditions such as airspeed, altitude, and geographic location, and their knowledge of the status of the aircraft, such as fuel level and the existence of or lack of mechanical malfunctions.

Environmental Protection and Noise Abatement

The public will continue to demand reductions in environmental contamination and reductions of acoustic noise over urban areas. As a result, additional environmental regulations or sanctions may be imposed on air carriers, and stricter noise control measures may be enacted. This will require that the United States collaborate with other nations in R&D focused on:

- propulsion systems that reduce emissions by utilizing alternative fuels, such as hydrogen, or hybrid fuels (one fuel for takeoff and another for cruising)
- processes (perhaps biologically or chemically based) to break down carbon dioxide into harmless components
- quieter engines and operations over urban areas, including revolutionary means to mitigate sonic boom effects over populated areas

Access to Space

Earth orbit provides opportunities for both civilian and military use of satellites for communications, navigation, and surveillance. It is anticipated that commercial firms and other nations will want on-demand access to Earth orbit that is quick and inexpensive. In most cases, sensors and communication devices are likely to be the dominant payloads, which will not require large spacecraft. Although low-Earth orbit was emphasized clearly by the workshop world teams, medium-Earth orbit and geostationary orbit will continue to be used for many satellite-based applications as well. Therefore, the steering committee has identified a need for systems and associated infrastructure to enable low-cost, on-demand delivery of small payloads with sensors or communications packages to any useful Earth orbit. The needs of future manned space activities and space science missions were not discussed.

NASA'S FUTURE ROLE IN MAINTAINING U.S. COMPETITIVENESS IN AERONAUTICS

The concept of "U.S. Aeronautics, Inc." provided the steering committee with a useful means of focusing workshop participants from government, industry, and academia on the accomplishment of a successful strategic planning workshop. Furthermore, the "future scenarios" planning methodology, although not the only way to conduct a strategic planning process, provided a means for searching out needs and opportunities and, ultimately, a basis for identifying technologies and other implications necessary to preserve options to deal with uncertainty 15 to 25 years in the future. Whatever the methodology, a way to provide a continuum of planning for uncertainty in a consistent and dynamic way is most useful if revisited on a periodic basis.

More important, however, the steering committee believes that the workshop and the concept of U.S. Aeronautics, Inc. represented a microcosm of a real partnership between government, industry, and academia. This partnership must continue to be fostered to achieve the goals outlined by the NSTC and, therefore, maintain the competitiveness of the U.S. aeronautics industry. Within this partnership, government must ensure that the conduct of long-term basic and applied research, the development of high-risk technology, the rapid validation of essential design and manufacturing tools and techniques, and the certification of products continue to be focused on these goals. The short-term, low-risk nature of most industry-sponsored R&D, despite the importance of long-term R&D, provides the partnership with no other viable alternative to a key federal role in maintaining the economic competitiveness of a market sector that is contributing favorably to the nation's balance of trade.[3]

Many options exist for continued government support of long-term aeronautics R&D. These include the rearrangement of current aeronautics R&D functions within the three agencies that currently carry out the majority of aeronautics R&D, which are NASA, the DOD, and the FAA; the assignment of responsibility to an existing government agency other than these three, such as the National Science Foundation (NSF) or the National Institute of Standards and Technology (NIST); or the creation of an entirely new federal agency. The elimination of an aeronautics program within NASA could be considered as part of any of these three options. However, the steering committee and workshop participants jointly agreed that it is not realistic to expect that government will be radically reinvented to respond to needs and opportunities represented by the five scenarios or, more generally, to the goals outlined by the NSTC. Therefore, the options listed above were rejected, and it was assumed that various agencies in the executive branch such as NASA, the DOD, the U.S. Department of Transportation (DOT), the FAA and the various committees of Congress will continue to share responsibility for government-funded aeronautics R&D. Given this assumption, the steering committee believes that within the federal government coordinated, cost-effective planning and implementation of long-term aeronautics R&D can only be accomplished by using the interagency process to designate a lead agency for this role. The steering committee further believes that NASA would best serve as the lead agency, rather than the DOD, the FAA, the NSF, or NIST, for the following reasons:

- NASA is chartered by the National Aeronautics and Space Act of 1958 to "preserve the role of the United States as a leader in aeronautical science and technology and the application thereof." No other federal agency has this legislative mandate.
- The NASA aeronautics enterprise has inherited its fundamental aeronautics R&D focus from its forerunner, the National Advisory Committee for

---

[3] In 1994, according to the NSTC, the aeronautics industry produced the largest trade surplus of any U.S. manufacturing industry, approximately $25 billion (NSTC, 1995).

Aeronautics, chartered in 1915. NASA has maintained this focus and has developed and maintained extensive R&D equipment and facilities. Although other federal agencies, such as the DOD and the FAA, also conduct aeronautics R&D and maintain appropriate facilities, this work is carried out in support of their operational missions. In contrast, the mission of NASA's aeronautics enterprise *is* aeronautics R&D.
- NASA has been charged by the Office of Management and Budget to develop an integrated national strategy and priorities assessment for civil aeronautics (NASA, 1995).
- NASA has responded to the goals outlined by the NSTC through its *Aeronautics Strategic Enterprise Plan for 1995–2000* (NASA, 1995). This plan includes a preliminary "road map" or strategic plan for the future of aviation that will be refined as a result of the workshop, this report, and the larger strategic planning process currently under way in the NASA Office of Aeronautics.
- NASA has several programs currently under way that already involve substantial partnerships between government, industry, and academia. These programs include the Advanced Subsonic Technology (AST) program, the High Speed Research (HSR) program, and the Advanced General Aviation Technology (AGATE) program.
- The future needs, opportunities, and implications for technology discussed in this report offered no compelling reason for the workshop participants or the steering committee to recommend an alternative to future NASA leadership, although the alternatives mentioned previously were considered.

Recommendation. To ensure coordinated, cost-effective planning and implementation of long-term aeronautics R&D within the federal government, the interagency process should be used to designate a lead agency for this role. The steering committee believes that NASA would best serve as the lead agency.

Leadership does not imply that NASA alone will have sufficient government funding to maintain the nation's global competitiveness in aeronautics. Nor does it imply that NASA will lead every R&D activity focused on the future aircraft, systems, and technology areas discussed in this report. It simply means that NASA should lead the government, industry, university partnership called for by the NSTC. Clearly, the DOD, the DOT, and the FAA will need to retain control of certain R&D initiatives that are pertinent to their day-to-day operations and national responsibilities. Existing interagency coordinating mechanisms, such as the Aeronautics and Astronautics Coordinating Board and the NASA/FAA Coordinating Committee,[4] also will continue to play an important role in ensuring effective R&D

---

[4] The Aeronautics and Astronautics Coordinating Board is co-chaired by the NASA Associate Deputy Administrator and the Undersecretary of Defense for Acquisition and Technology. The NASA/FAA Coordinating Committee is co-chaired by the NASA Associate Administrator for Aeronautics and the FAA Associate Administrator for Research and Acquisition.

coordination. However, the steering committee believes that NASA can provide effective national leadership in maintaining the superiority and competitiveness of U.S. aeronautics through a renewed emphasis on long-term R&D.[5] This is particularly true for those areas of research and technology that could have a direct impact on the civil marketplace and in those leading-edge generic research and technology areas that would lead to either military or civilian applications. The same would be true for long-term, high-risk research and technology areas that intersect with FAA responsibilities that relate to aircraft safety, infrastructure efficiency, and environmental impact. As recommended in *Aeronautical Technologies for the Twenty-First Century*, NASA should take the leadership role in high-risk/high-payoff research related to these areas of shared responsibility (NRC, 1992).

An in-depth assessment of the specific programs and long-term R&D activities that NASA should engage in as the lead agency for aeronautics is the next logical step in this current strategic planning process. In addition, the roles of other federal agencies, private sector organizations, and academic institutions that are part of the nation's aeronautics partnership must be carefully considered and defined. The steering committee believes that this next phase of the strategic planning process should again be conducted with broad participation from government, industry, and academia and should proceed without delay.

REFERENCES

NASA (National Aeronautics and Space Administration). 1995. Achieving Aeronautics Leadership: Aeronautics Strategic Enterprise Plan, 1995–2000. Washington, D.C.: National Aeronautics and Space Administration.

NRC (National Research Council). 1992. Aeronautical Technologies for the Twenty-First Century. Aeronautics and Space Engineering Board, Committee on Aeronautical Technologies. Washington, D.C.: National Academy Press.

NSTC (National Science and Technology Council). 1995. Goals for a National Partnership in Aeronautics Research and Technology. Executive Office of the President, Office of Science and Technology Policy. Washington, D.C.: National Science and Technology Council.

---

[5] The steering committee envisions that NASA's role in the development of technology would not extend beyond what is referred to by the DOD as 6.3A—Advanced Development.

APPENDICES

APPENDIX A

Statement of Task

The Workshop Steering Committee will participate in the development of long-term global aeronautics scenarios to help guide the strategic planning process for the NASA Office of Aeronautics. These scenarios will span the breadth of aeronautics, including civil aviation, military aviation, and access to space. In addition, the scenarios will consider other critical factors that will impact the future of aeronautics, including information and communications systems, national and global transportation systems, and air traffic management systems, as well as economic, social, and policy factors. These scenarios will focus on the long-term, which is defined as 15 to 25 years or beyond the next-generation systems.

The scenarios will draw on existing data and studies, including research in this area prepared under NASA auspices by SAIC (Science Applications International Corporation). Scenarios based on revolutionary, as well as evolutionary, technologies will be considered. The workshop will further refine and develop the scenarios by reviewing the key driving forces in aeronautics and will strive to reach consensus regarding the key issues, the most likely scenarios, and the general role that NASA should play in planning for the future.

APPENDIX B

Biographical Sketches of Steering Committee Members

William W. Hoover (chair) is the former executive vice president of the Air Transport Association of America and is a retired U.S. Air Force major general. At the Air Transport Association, he was responsible for all aspects of the association's activities, including development and implementation of wide-ranging airline policies. While on active duty, Gen. Hoover spent four years in the U.S. Air Force space program, was a combat air wing commander in Vietnam, and later served as deputy assistant secretary for military applications in the U.S. Department of Energy. Upon retiring from the Air Force, he served as assistant secretary, Defense Programs, U.S. Department of Energy, and was responsible for the U.S. Nuclear Weapons Program. He has a B.S. degree in engineering from the U.S. Naval Academy and an M.S. degree in aeronautical engineering from the Air Force Institute of Technology.

Guion S. Bluford, Jr., is vice president and general manager of the Engineering Services Division of NYMA, Inc. As a retired colonel in the U.S. Air Force and a former astronaut, he has an extensive aeronautics background. Operational experience includes 65 combat missions over North Vietnam, duties as an Air Force instructor pilot, and four space flights as a shuttle mission specialist. Dr. Bluford's research activities include computational fluid dynamic studies of advanced aeronautical concepts at the Air Force Flight Dynamics Laboratory. He has a B.S. degree in aerospace engineering from Pennsylvania State University and M.S. and Ph.D. degrees in aerospace engineering from the Air Force Institute of Technology.

Richard S. Golaszewski is executive vice president of GRA, Inc., where he specializes in aviation economics, safety, and public policy and studies the economics of airports, airlines, and aircraft manufacturing and safety. Before joining GRA, Inc. as an economist in 1977, he was a lecturer at La Salle College, where he received a B.S. degree in accounting, and a lecturer at the Wharton School of the University of Pennsylvania, where he received an M.P.A. degree. Mr. Golaszewski is a member of the Aviation Economics and Forecasting Committee of the National Research Council Transportation Research Board, the Economics Technical Committee of the American Institute of Aeronautics and Astronautics, the American Helicopter Society, the Air Traffic Control Association, and is aviation editor of the *Journal of the Transportation Research Forum*.

William H. Heiser has extensive industrial engineering and management experience with Pratt & Whitney Aircraft, General Electric, and Aerojet General. In the 1970s, he was chief scientist at the Air Force Aeropropulsion Laboratory and the Arnold Engineering Development Center. Dr. Heiser has been professor of aeronautics at the U.S. Air Force Academy since 1989, a member of the Air Force Scientific Advisory Board since 1992, and chairman of the U.S. Air Force Pratt & Whitney F-119 Executive Independent Review Team since 1993. He has a B.S. degree in mechanical engineering from the Cooper Union, an M.S. degree in mechanical engineering from the California Institute of Technology, and a Ph.D. degree from the Massachusetts Institute of Technology. He is also a fellow of the American Association for the Advancement of Science, American Society of Mechanical Engineers, and American Institute of Aeronautics and Astronautics.

Grace M. Robertson is vice president–general manager, Developmental Programs for the Douglas Aircraft Company. Prior to joining this company in 1994, Ms. Robertson spent 17 years at Boeing, moving up in management after beginning as an avionics design engineer. She received a B.S. degree in electrical engineering from the University of Wyoming and an M.S. degree in management from Stanford University, where she was selected as a Sloan fellow.

Jeffrey K. Schweitzer is manager, Conceptual Design & Systems Engineering, United Technologies/Pratt & Whitney. In this position he is responsible for the conceptual design definition of advanced gas turbine and rocket propulsion systems and the long-range technology planning required to support future product and business strategies. Mr. Schweitzer began his career with Pratt & Whitney as an experimental engineer in 1972. He has a B.S. degree in aerospace engineering from Pennsylvania State University.

Thomas B. Sheridan is a professor of aeronautics and astronautics and a professor of engineering and applied psychology at the Massachusetts Institute of Technology. He is a former president of the Human Factors Society and a fellow of the Institute of Electrical and Electronics Engineers. Dr. Sheridan is a member of the National Academy of Engineering and is currently serving on the National Research Council Panel on Human Factors in Air Traffic Control Automation.

Robert E. Spitzer is currently vice president for engineering of the Boeing Commercial Airplane Group. In this position he leads company efforts to gain enabling technologies and maintain technical excellence and is responsible for a broad range of research and development-related activities. Mr. Spitzer has been an employee of Boeing since 1965 and has worked in a number of military and commercial research and development programs. He has engineering degrees from the University of Illinois and the California Institute of Technology.

APPENDIX C

Workshop Participants

William W. Hoover
Chair
Maj. Gen. U.S. Air Force (retired)
Williamsburg, Virginia

Leonhard Allgaier
Leadership Processes, Systems and
  Technologies
West Bloomfield, Michigan

Howard Aylesworth, Jr.
Director, Airworthiness and Regulation
Aerospace Industries Association of
  America
Washington, D.C.

Guion S. Bluford
Vice-President–General Manager
NYMA, Inc.
Brook Park, Ohio

Joseph Breen
Transportation Research Board
National Research Council
Washington, D.C.

Donald J. Campbell
Director
NASA Lewis Research Center
Cleveland, Ohio

Richard S. Christiansen
Aerospace Research Division
NASA Office of Aeronautics
Washington, D.C.

Jerry Creedon
Director
NASA Langley Research Center
Hampton, Virginia

William Dean
Deputy Director
NASA Ames Research Center
Moffett Field, California

Thomas DuBell
Director
Component Center (Combustor
  Augmentor Nozzle)
Pratt & Whitney,
Jupiter, Florida

Alan H. Epstein
Department of Aeronautics and
  Astronautics
Massachusetts Institute of Technology
Cambridge, Massachusetts

Angela Gittens
Aviation General Manager
Hartsfield Atlanta International Airport
Atlanta, Georgia

Richard S. Golaszewski
Executive Vice President
GRA, Inc.
Jenkintown, Pennsylvania

Vicki Golich
Associate Professor of Political Science
California State University, San Marcos
San Marcos, California

James M. Guyette
Former Executive Vice President,
　United Airlines
McAllister, Montana

William H. Heiser
Professor of Aeronautics
U.S. Air Force Academy
USAF Academy, Colorado

Michael Henderson
Program Manager,
High Speed Civil Transport
Boeing Commercial Airplane Group
Seattle, Washington

Ned Hogan
RADM U.S. Navy (retired)
Sun Valley, Idaho

Lee Holcomb
Aviation System Technology Division
NASA Office of Aeronautics
Washington, D.C.

John J. Kelly, Jr.
Brig. Gen. U.S. Air Force (retired)
Centreville, Virginia

Doug Konitzer
Staff Engineer
GE Aircraft Engines
Cincinnati, Ohio

Robert G. Loewy
Chair, School of Aerospace
　Engineering
Georgia Institute of Technology
Atlanta, Georgia

Douglas K. Mang
Director, Raytheon Aircraft Programs
Raytheon Company
Arlington, Virginia

James W. Mar
Professor Emeritus
Massachusetts Institute of Technology
Pacific Grove, California

Gene McCall
DOD-PO
Los Alamos National Laboratory
Los Alamos, New Mexico

Harry McDonald
Director
NASA Ames Research Center
Moffett Field, California

O. Wayne McGregor, Jr.
Director, Technology Development
Lockheed-Martin Tactical Aircraft
　Systems
Ft. Worth, Texas

Duane T. McRuer
Chairman
Systems Technology Inc.
Manhattan Beach, California

Terry Neighbor
Air Force Material Command, STX
Wright Patterson AFB, Ohio

Robert B. Ormsby, Jr.
Aircraft Group President (retired)
Lockheed Corporation
Newhall, California

Appendix C

Clinton V. Oster, Jr.
School of Public and Environmental
  Affairs
Indiana University
Bloomington, Indiana

David Plavin
President
Airports Council International of
  North America
Washington, D.C.

Michael J. Prather
Head, Department of Earth System
  Science
University of California
Irvine, California

George Price
Director, Engineering Advanced
  Projects
Sikorsky Aircraft
Stratford, Connecticut

Grace M. Robertson
Vice-President–General Manager
Design and Technology
MD-90 Program Manager
Douglas Aircraft Company
Long Beach, California

Jeffrey K. Schweitzer
Manager
Conceptual Design Systems
  Engineering
United Technologies/Pratt & Whitney
West Palm Beach, Florida

A. Richard Seebass
Professor and Chair of Aerospace
  Engineering Sciences
University of Colorado
Boulder, Colorado

Thomas D. Sheridan
Ford Professor of Engineering and
  Applied Psychology, Professor of
  Aeronautics & Astronautics
Massachusetts Institute of Technology
Cambridge, Massachusetts

Robert Simpson
Flight Transportation Associates
Cambridge, Massachusetts

Agam N. Sinha
Director, Air Traffic Management
  Division
The MITRE Corporation
McLean, Virginia

Robert E. Spitzer
Vice President, Engineering
Boeing Commercial Airplane Group
Seattle, Washington

Ken Szalai
Director
NASA Dryden Flight Research Center
Edwards, California

William Thurman
Lt. Gen. U.S. Air Force (retired)
Pinehurst, North Carolina

Robert Whitehead
Associate Administrator
NASA Office of Aeronautics
Washington, D.C.

Henry Winkler
Vice President, Special Projects
Hughes Information Technical
  Systems
Fullerton, California

Andres Zellweger
Director, Office of Aviation Research
Federal Aviation Administration
Washington, D.C.

NASA Core Team Personnel

James Afarin
Doug Dwoyer
Joe Elliott
Peter Ouzts
Robert Pearce
Frederick Schmitz
Jim Stewart
Howard Wesoky

The Futures Group/SAIC

Robert Avila
Harry Gehring
Peter Kennedy
David Louscher
Lee Lunsford
Paul Rich
Ken Sawka
Charles Thomas

ASEB Staff

David A. Turner
Study Director

Victoria P. Friedensen
Senior Program Assistant

APPENDIX D

Scenario Narratives as Provided by the NASA/TFG/SAIC Core Team

Pushing the Envelope

U.S. Economic Competitiveness	Worldwide Demand for Aero Products/Services	Threats to Global Security and/or Quality of Life	Global Trend in Govt. Participation in Society
Strong	High Growth	Low	Low

Summary

The malaise of the late twentieth and early twenty-first century seems like an unpleasant but distant memory. It is 2015 and the U.S., Western Europe, and Japan have reasserted their economic dominance over a world in which free trade, open markets, and democratic government (broadly speaking) reign supreme. A rising global economic tide is gathering many, if not all, ships, even former "basket cases" in the developing world and the former Soviet Union. The World Trade Organization (WTO) and other multilateral organizations keep markets open and succeed in maintaining favorable conditions for cross-border commerce and finance. The information revolution is in a new, dynamic phase and corporations that have learned to leverage this power invariably lead their respective industries. Society is highly mobile, within the bounds of socioeconomic class and national origins. Highly skilled professionals enjoy extraordinary geographic mobility.

The world is thankfully free of large-scale military conflicts. International rivalries tend to be played out in boardrooms and labs, not military battlefields. Competition is intense, with generally low entry barriers and ample venture capital available for promising new ventures. Consumers possess unprecedented opportunities for financial growth, but are time-impoverished and weary in this highly competitive, pressure-cooker world.

The Futures Group

The pessimists had it wrong. Those dreary, end-of-the-millennium predictions of the U.S.'s (and Europe's) demise *were* greatly exaggerated. The West was neither dead nor dying. It *was* troubled, economically, politically and socially. But, as events eventually proved, these problems were not intractable; we would overcome them and rebound with a fury. In fact, notwithstanding our huge debt, aging infrastructure, and slumping wages, the U.S. and Western Europe were rebuilding themselves in a way that would ensure economic growth, income recovery, and technological superiority far into the twenty-first century.

But I am getting ahead of myself. In 2000, it was easy to overlook our underlying competitive strengths. The U.S., as well as Western Europe and Japan, had not yet come to grips with the big problems confronting society, particularly those related to debt, long-term funding of entitlement programs, and unemployment. Existing programs were expensive, rife with inefficiencies, and fiscally insupportable. Political systems were mired in what we call "gridlock" with politicians pursuing votes via painless "solutions" (like restricting immigration, raising minimum wages, etc.) that were at best irrelevant, and often counterproductive. Markets were edgy. Wall Street looked no further than the next quarter's earnings, knowing too well that the status quo was unsustainable.

The twentieth century ended on a troubled international note. Concern centered mostly on Asia. Japan experienced an acute financial crisis and near total economic meltdown that would require several years of debt write-offs and bank restructurings to correct. Substantial growth pains afflicted China. The nation was forced to come to grips with the realities of a globally integrated marketplace in which its participation was desired, but not required. Moreover, its rapid economic growth and modernization increased the flow of knowledge throughout society, fueling discontent especially in urban areas over limits to freedom and political action. Authorities chose stability over democracy. Consequently, China moved toward a form of authoritarian capitalism, which combined highly centralized government control over an increasingly decentralized and private economy.

Meanwhile, as the new century was launched, the Emerging Markets were doing generally well. Korea was, in effect, a developed market and within a few years would be admitted into the Organization of Economic Cooperation and Development. India, Brazil, and Indonesia were among the fastest-growing, most dynamic economies anywhere, with rapidly expanding middle classes and affluent populations. Not all were doing equally well, however. Again, China was an inconsistent performer and investors feared that it could suffer political turmoil and a major economic setback. The Middle East was still politically unstable, leaving global energy markets constantly on edge.

Japan's financial market meltdown in 2001 was a crystallizing event — a rallying cry for visionaries who had long argued for fundamental reform. Had the Group of Ten (G-10) central bankers not offered immediate and effective support, Japan's economy would have plunged into a long, deep recession or depression — probably taking the rest of the major markets down with it. But coordinated G-10 action in the form of instant liquidity and commitments of long-term credits staved off a disaster. All was not well, however. Even as Japan's situation began to stabilize, global currency and equity markets were rocked by near-panic buying and selling. More extensive G-10 central bank cooperation ensued with the objective of achieving coordinated stabilization of key currencies. To make this work, all the major countries had to agree to some fairly aggressive fiscal targets.

By 2002, the plan was in place and most of the G-10 stuck to their commitments. Politicians in effect told their respective constituencies, "Look, do this, or we'll end up like Japan, or worse." Most went along, however grudgingly. Constituents deeply feared what would happen if strong medicine were *not* swallowed. And to a point they were willing to follow the politicians.

The results were startling and far more positive than anyone could have imagined, especially for the U.S.. The Chief Executive and Congress hammered out a far-reaching restructuring of government spending programs, including entitlements like Social Security. The idea was to seize this window of opportunity while it was still possible and restructure the entire business of government in the most politically neutral, socially responsible way possible.

The process was anything but silky smooth. Medicare reform was predictably a political minefield. But in general most everyone perceived the process as fair. Everyone had to give; few gained outright. There were very few loopholes in which the wealthy could take shelter. Consumer advocates kept the lobbyists at bay and the politicians reasonably well behaved. Privatization attenuated economic dislocation, as it provided instant revenue and therefore meant lighter spending cuts. Moreover privatization proceeds covered the investment needed in critical infrastructure, including satellite communications and fiber-optic backbone. These were pitched as exceptional activities, actually, designed to "jump start" markets before fully liberalizing them via privatization.

A mix of market and fiscal incentives helped the public swallow the harsh reform medicine. (In truth, subtle anti-rich appeals helped a lot, too, for the wealthy had the most to lose from the proposals being debated). The tax structure rewarded savings and penalized frivolous consumption. Tax credits were granted for eldercare at home. Various user fees were reviewed and, if necessary, changed to ensure that fees for government goods and services reflected true costs to the public. Conservatives' dream of school coupons succeeded in not only providing choice but also quality education for middle income and (most) children from disadvantaged backgrounds. In major U.S. cities, much public education was effectively outsourced with the help of Catholic and other religious schools.

In retrospect, the sum effect of these major policy shifts would have been limited had U.S. industry not been primed to exploit the improved business environment. In fact, the much maligned "reengineering" and "restructuring" trends of the 1990s proved to be the boot camp of the emerging global business environment. By the late part of the decade, productivity gains were appearing in spades, not only because work forces were slashed to the bone, but because companies were finally growing skilled in applying information technology to all phases of the value chain, from design to distribution. Each wave of new workers was more and more inclined to embrace and innovate information solutions.

Naturally, the Internet was an extraordinary enabler of this "takeoff." Its impact by 1998 was clear even without the rather significant broadband investment that took place in the U.S. and then later in Germany, England, and France after 2000. By 2005, nearly every U.S. household and commercial center was fiber optically connected. The same was true of Europe (by 2010). In both instances, coordinated government-industry investment engendered relatively efficient and harmonious development, with ample commercial space for private entrepreneurial activity and, of course, profits.

These broadband development projects were exceptional in the sense that they were *not* indicative of a more general move to industrial policies in the U.S. or anywhere. Even socially liberal politicians did not advocate a return to generous welfare policies or state capitalism. Yet, at the same time, there was broad recognition that without the government setting the rules and directing the information infrastructure that either the big, capital intensive work would never get done or it would be badly done, in staccato, and probably chaotic fashion.

By 2005, the market-based restructuring of the U.S. economy was nearly complete. Japan was deep into its own post-crisis restructuring program and the major European countries had their own programs under way. The success of the international organizations during the post-Japan currency scare greatly increased the stock of the WTO. In fact, the WTO played a critical role when the major trading nations could easily have turned protectionist. The WTO not only staved off protectionism, it also accelerated market liberalization as the major trading nations got firmly on their feet.

IT and telecom innovation and expansion accelerated after 2005, as bountiful venture capital fed countless start-up companies offering new products and services for constantly evolving and changing technologies. Businesses ever hungry for efficiency enhancements kept demand high; so did households seeking the latest and greatest interactive educational and entertainment software that broadband communications could provide. With nearly all U.S. households wired by 2010, two in five white- or pink-collar workers now work from home full time; many more do so part-time.

High levels of global trade drove the development of globally harmonized product standards. Trade itself was global and multilateral in focus. Regional trade groups were not really trade blocs at all; they were more like halfway houses to global trade for newly reforming countries. By 2010, only a handful of significant trading nations had significant tariff and nontariff barriers. Offsets were now second generation — for example, Vietnam requiring them of Chinese companies wanting to sell aviation equipment locally. China pared back its own offset requirements; with its booming economy, China no longer needed them.

Free-trade conditions and the Internet facilitated technology transfer to the new manufacturing and knowledge-based economic zones. Workers on multiple continents were now able to work closely on common (design, engineering, for example) projects. Language and time zone differences were trivial complications to these cross-national, cross-cultural collaborative efforts. U.S. and European-based companies discovered that creating cross-national development teams not only was attractive from a cost standpoint but also ensured high levels of local intellectual content — so critical for a truly global marketplace.

The WTO evolved into an increasingly important international organization, respected by mature and developing nations alike for its fairness and efficient bureaucracy. The IMF became more involved with solvency problems of very poor countries, while the World Bank actively supported free-market infrastructure projects via investment credits and small, short-term equity positions in start-up operations. Private investment continued to be the driving force of infrastructure, however, among other things thanks to highly dynamic venture capital markets and ever-sophisticated risk management tools.

The other major surviving international organization was the UN, whose traditional activities were pruned but continued nonetheless to supply a structure for international peacekeeping and humanitarian relief activities. The U.S. and Europe contributed high-tech weaponry to these police activities. Across the world, the U.S. military presence was significantly downsized and limited primarily to a residual presence in Southeast Asia and the Middle East, which remained subject to low-level instability. Globally, however, it was a time of low weaponry demand.

Free trade and globalization had profound impacts on highly skilled human capital, which could move comparatively easy across borders as major trading nations embraced liberal immigration codes if they enhanced national competitive advantage. Scientists from developing countries trained in the U.S. faced difficult choices when their visas expired: Stay and live reasonably comfortably in the U.S. or go home and make perhaps a small fortune.

The battlefield for power and influence in the world was no longer military but economic in nature. Only small, economically insignificant nations worked out their differences by taking to arms. For all other nations, power resided in the sum of their competitive economic advantages vis-à-vis their trading partners. This was not exactly just "good sport"; competition was typically cutthroat, in the extreme bordering on unethical and even unlawful. For the most part, the major economic powers maintained high intellectual property protection but below them there was widespread piracy and ripping off. The economic cost to the inventing nations was mitigated by the sheer speed of product innovation and the fleetingness of product life cycles.

Nationalism itself was undermined in the global, information-driven economy through a combination of national-to-local "downloading" of funding and program responsibility, and because of the radical decentralization of institutions brought about by the Internet and advanced communications technologies. As goods, services, capital, and in many cases people flowed fluidly across national borders, subnational and cross-national economic zones evolved into the most important sources of economic growth. These New Economic Geographies (NEGs) lacked a formal, juridical identity, but they were real and increasingly important poles of new investment, technological innovation, and job creation. Here in North America, Silicon Valley was a high-tech precursor of the NEG trend. By 2010, there would be NEGs such as the San Diego-Tijuana manufacturing belt, the South Florida-Caribbean commercial area, and the Southeastern Brazil industrial (incorporating Northeastern Argentina) among dozens of others throughout the world. In the developing world, an important demographic result was the redirection of rural emigrants away from old, heavily congested urban areas, toward these booming NEGs. These NEGs required massive amounts of both hard (roads, bridges, rails, airports, power, water, etc.) and soft (housing, hospitals, schools, etc.) infrastructure investment. More than half of these projects were financed through build-operate-own/transfer-type privatization schemes.

Consumers in the mature markets of the U.S., Europe, and Japan are economically healthy for the most part, but not secure. Paradoxically, there is little economic security in this strong, dynamic economic environment. Affluence is widespread but requires enormous time commitment to work and constant skill upgrading. Competition in the workplace is intense and unrelenting. Even in Europe, two-week vacations have become extremely rare; there is money but a paucity of time.

Adding to time poverty, middle- and upper-class mature market consumers (which now count Taiwan, Singapore and Korea among their ranks) manage their own pension, insurance, and benefit programs. The plethora of international investment opportunities has made this much like a potentially lucrative sport — a very addicting one at that. In the absence of government or pension programs, savings rates are high as consumers realize they themselves must provide their own next retirement funding, with bare-bones government help.

Mature market consumers (and to some extent professional/affluent classes in Emerging Markets) are environmentally aware, with strong quality-of-life feelings. Life is demanding, time is short, and few will tolerate having their precious little leisure time ruined by dirty air, polluted lakes, and noisy skies. At the international level, the United Nations Organization for Protecting Environmental Resources (UNOPER) manages a voluntary program in which member nations buy, sell, and barter CO_2 "pollution rights." Pollution "credits" are allocated on the basis of industrial development and per capita income. It has the effect of discouraging highly polluting industries from producing in the developed world.

Within mature market countries, market incentives take the form of privatized roadways and bridges, as well as "smart highways" in which automobile operators are charged differential rates for use of highways and bridges. This turns out to be a lucrative new business opportunity. A wide range of other market-based solutions are effectively applied as well. Meanwhile, the scope of the Environmental Resources Management Agency (formerly the Environmental Protection Agency) is radically reduced and different. ERMA now polices a much narrower range of environmental issues — groundwater pollution, for example, and others not handled by the market.

The environmentalism is not ideological; it is personal (some say selfish) and practical in nature. Likewise, mature market consumers are concerned about infectious diseases, which the World Health Organization (WHO) data prove have been on the sharp upswing as many of the more remote developing world "hot zones" have been brought into contact (via commerce, travel, missionary and scientific work) with the more developed population areas. WHO has expanded surveillance of Ebola-type outbreaks in response to several major crises in Africa, South Asia, and even Europe.

Natural and man-made health threats encourage affluent consumers to live far from the sources of these problems. Communication technology allows many to live in safe, environmentally clean planned communities where urban and suburban problems do not have to be faced.

Meanwhile, in the Emerging Markets, rapid economic growth and heightened consumption (gasoline, petroleum products, meat, etc.) have rendered natural resources (especially air and water) extremely vulnerable to contamination or depletion. With limited success, some Emerging Market elites and intellectuals pressure their respective governments to join the global Green movement by embracing high (if not First World) standards of environmental stewardship.

Rapid economic expansion, political reform, and liberal global trade combine to accelerate the growth of Emerging Market middle and affluent classes. Consumption potential is enhanced by the deepening of financial markets and the availability of consumer credit, courtesy of new financial market players such as Citicorp, GE Capital, and Household Finance. The ranks of first-time car buyers expand tremendously. Emerging Market consumers show no loss of appetite for all kinds of consumer goods. And increasingly, especially in Asia, the new middle classes are discovering the wonders of Europe and the Americas. Hundreds of local travel companies now organize tours of London, Paris, Rome, and New York, among other prime destinations.

Scenario Matrix - Pushing the Envelope

Scenarios / Drivers	Pushing the Envelope
World Economy and Market Environment	Dynamic, global economy with highly integrated markets; overall finance stability at macro level, volatility at enterprise level; capital is available but risk adjusted cost of capital; high demand from mature and emerging markets; R&D = small r, Big D; short termism; a large affluent class, growing middle class world-wide; overall low unemployment and flexible labor markets (telecommuting at high end labor).
International Trade Environment	World Trade Organization kind of world establishes and enforces free and fair rules of the global trade game; second generation offsets between emerged & emerging markets (e.g., China & Vietnam); few residual regional preferences, EU, NAFTA; harmonization on navigation, environmental via strong ICAO; blurring of air line nationality (e.g., AA/BA/USAIR); market determines landing rights, with congestion based pricing in mature markets; privatization of ATC in mature and select emerging markets; development of internet based open market for airline seats; airlines move to leased aircraft and provision of air services.
Political Instability	Fairly Stable; some low level instability in Middle East (religious / ethnic) and Russia (little border tension); social tension between haves and have-nots.
U. S. Military Requirements	Low level US "police" role under UN auspices, US and Europe contribute high-tech weapon systems to police force; some US military presence in Middle East; problems are local bullies, civil strife; significant reduced basing around world; European unification limits need for strong NATO; some basing in South East Asia; rapid deployment forces; low level of global conflict leads to low armaments demand.
Global Distribution of Power & Technology	Increased number of effective actors; high level of rapid technology diffusion fed by global infonet; still have and have not states and groups; many niche economic powers; information security is problem, particularly in commerce.
Fuels & Fuel Sources	Oil is critical and some Middle East instability; plus

Appendix D

Scenarios Drivers	Pushing the Envelope
	environment quality issues lead to alternate fuels; market driven access to resources.
US Policy	Basic policy - laissez fair, (in both economic and social policy); tort reform combined with strictly enforced truth in labeling laws; anti-trust laws much more liberally interpreted due to dynamic high-tech information rich nature of economy; bankruptcy laws strictly enforced; tendencies to off-load social responsibilities to state and local governments; minimal government, privatization of services; no deficit/reduced debt, tendency not to subsidize industry; minimum body of law and regulation to support a dynamic free market economy; affluent charities for humanitarian work.
Corporate Structure and Operations	Trend to virtual, global decentralized structures, with mobile contingent labor; technology proliferation is high and dynamic; market share is dynamic, in general very hard to capture share for a long time; 24 hour operations; significantly reduced offsets (some still exist in excess capacity segments or in newly industrialized countries); alliances very important, but flexible, often short term; ownership patterns are dynamic with variable mix of capital sources; corporate governance is often global and virtual; labor competition is intense at high end, but less so at the low end; trend toward converging wage rates world wide.
Environment	This is key quality of life issue; attitudes and solutions to pollution vary across world ranging from don't care to highly engaged in prevention and clean up; often back yard issue that plays out in politics; generally this world is moving toward "green"; as a more affluent society the US is pushing market based environmentalism and marketing green technology world wide.
Public Health	General levels of public health are high and follow quality of life; very serious spikes in global infectious diseases due to high mobility; environmental epidemeology is well funded.
Public Attitude to Technology	Love that stuff; "anywhere, anytime" newness, whimsy, fads; seen as good, as a solution rather than the problem; popular support for "big science".
Education	Global meritocracy, in love with education; education is life long not in discrete chunks; multiple careers; in "sciences" its technical more than basic research; focus on business and

Scenarios / Drivers	Pushing the Envelope
	culture and international business standards; very global and interconnected, some virtual schools; education also is a quality of life activity; funding is from all sources and very high; still an underclass with inadequate education; world class graduate education in the US; because of access to information many people have higher knowledge and interest in many areas including science.
Geographic (Living) Dispersion	More and more people live where they want across the globe, not tied to work location (telecommuting); reduced population in some old urban and suburban environment; still high business travel, but now from highly dispersed locations.
Communications and Information Technology	This is the backbone of the world; it is ubiquitous and the heart of global business and personal life; high bandwidth, mobile, lots of new hardware and application (much of it is useless) very high competition; a lot of turmoil in products and marketing; communication and information security important for personal, business and national security.
Production Cost Performance	Many product components are commodities, but value-added is created through rapid small lot customization and/or very high-tech non-commodity components; business cycles tend to be globally synchronized; ownership patterns are very flexible (time shares, multi-vehicle leases, etc.), very disbursed ownership patterns for transportation products; barriers to entry/exit are very low; many products experience rapid obsolescence.
Technology development and Application	Very, very market driven; global financing instruments are key to R&D (with emphasis on "D"); quality of life drives many new technologies; consumers like technology; short term product environment has low market incentives for R&D with long pay back period; long-term research comes from high risk capital investment and portfolio management but rapid growth environments create new opportunities and challenges.
Time Poverty Leisure Time, Entertainment	Very time constrained in dynamic global economy, but separation of work and play is unclear; a lot of contract labor-work for a while, take a long vacation (implies excellent personal credit management); blurring of education and leisure; options for leisure are open ended; middle class tries to mimic this activity, but with fewer resources; entertainment is huge industry and has expanded

Scenarios / Drivers	Pushing the Envelope
	definition (TV, computer, internet interactive combinations with personalization and "preference buffers").
Global Transportation Infrastructure	Trend toward privatization of development, ownership, and operation of infrastructure; "smart" (user-friendly) infrastructure; harmonization of standards especially in automation and communications; pay for access - user fees; infrastructure must accommodate low-cost, rapid-time-to-market, optimized production flows; potential for electro-magnetic/ frequency congestion; supports rapid response to consumer demand (internet orders/overnight delivery); dynamic mix of transportation modes (e.g., "transportation credit cards/super transport rental company"); high-end market for significantly reduced portal-to-portal travel time and for continuous connectivity.
Safety and Security	Safe global transportation system driven by infonet informed consumers; must be globally harmonized/standardized (ICAO); security threats including data are widely varied and very hard to predict; data security is important because of smart infrastructure.
Access to space	If space is the most cost effective way to solve a problem, global financing will be possible, but, it is likely to be private and entrepreneurial; (note: small "r", BIG "D"); public "big science" programs include continued human and robotic exploration and exploitation of space (possible public/private mix, (e.g., advertising, entertainment)).

Grounded

U.S. Economic Competitiveness	Worldwide Demand for Aero Products/Services	Threats to Global Security and/or Quality of Life	Global Trend in Govt. Participation in Society
Strong	Low Growth	High	High

Summary

Out of dreadful necessity, the world has replaced "being there" with virtuality. For too many people in too many places the rate and pace of change stripped them of their humanity. Angry, frustrated, bitter, jealous individuals and groups with increasing frequency resorted to unprecedented and unpredictable violence against large masses of people wherever they could be targeted. All forms of mass transit, large public gatherings, dense urban environment (despite onerous and very expensive security measures) are now avoided, if possible. The global economy has, however, avoided collapse and indeed more than risen to the challenge thanks to the infocom revolution. Travel may be but a specter of its past, but massively expanded bandwidth now ties the world far closer than it ever was in the past. All aspects of life, commerce, government, culture, and entertainment flow continuously and globally through the G-net. The economy's vitality is still derived from the rapid pace of technological advance, product innovation, and entrepreneurship. Goods still move great distances through the global economy, occasionally at risk, but goods are expendable; people move only when absolutely necessary. Each day it becomes more amazing what simulation can do.

The Futures Group

Global Business in 2020. (New York: Harcourt Brace and Pearce, 2021.) An introductory essay to the second edition.

No matter how elegant are the arguments that "technologies emerge to match the need," there is no getting around the common-sense position that, as a civilization, we have been very lucky. If the terrorplots and mass killings of the mid-2000s had swept through the globe a few decades earlier, democracies might not have survived the need for oppressive police controls to keep societies functioning. Beyond that, the growth of the global free market (and the improvements in worldwide living standards that have ensued) might have been postponed indefinitely. It is unlikely that the more fragile technological and economic systems of that earlier era could have withstood the onslaught of death, terror, and infrastructure destruction that we endured just a few short years ago.

As an interesting (but relevant) example, many in that peaceful and complacent era (1960 through 2005) would have said that it is implausible (perhaps impossible) that a global economy and society could thrive without dependence on the rapid growth of air transport. It was one of those conventional wisdoms — the growth of the global economy and the growth of global aeronautics were synonymous. They were mutually dependent on each other. As the world economy began a serious phase of rapid globalization in the 1990s and early 2000s, it was common, even in periods of economic recession, to forecast the symbiotic growth of aeronautics. It was so obvious. What could stop or replace its central role, especially in the high-growth emerging markets?

This textbook has no history section; and, while I am certain that all readers are familiar with the events of the past twenty years, I wish you to immerse yourselves, briefly, in the mindset of optimism and growth that characterized the late 1990s and early 2000s. It will give you an important perspective on today's business and political climate. It should also teach you never to assume; never assume you can predict even the most "obvious" trends.

The world economy of the late 1990s and early 2000s was growing at unprecedented rates. Moreover, it was a growth pattern characterized by interconnectivity of businesses, organizations, and individuals on a worldwide scale. That dynamic growth and interconnectivity sat on six pillars — the U.S. economy as an early engine of growth, relative global peace, free trade, telecommunications, fast and reliable transportation, and efficient global capital markets. While the mature market economies of North America, Europe, and Japan grew steadily, they pumped capital into the rapidly developing emerging markets of Southeast Asia (SEA) and Latin America. Products were manufactured and assembled on a global scale and sold into the mature affluence of the Northern industrial powers and into the burgeoning middle classes of the emerging markets.

The U.S. economy did well during this period and the government spent some of its wealth putting its economic and social house in order. While industries globalized, consolidated, and spun off multiple service sector support firms, the government used the increased revenues from corporate and income taxes to end the deficit and pay down the debt substantially. If the society was spending too little on infrastructure and alternatives to conventional technologies, ... well, those issues would either respond to market forces or be dealt with in the coming years. The U.S. shifted farther into a modestly regulated economy with market incentives for enforcement. While some serious "transition" inequities arose in social services,[1] the evolution of safety-net spending to the states went relatively smoothly. This was a time of global business interests and minimalist government enabled by a fairly peaceful world.

Europe and Japan continued to lag somewhat under the burden of higher and more intractable public debt, but nevertheless took part in the globalizing economy. They, too, saw their industries move manufacturing abroad, expanding to the promise of rapidly growing markets, while profits flowed back home. Like U.S. industries, the Europeans tended to invest in all regions, while the Japanese had a tendency to focus on SEA. This is a good time, however, to point out that it was not so much the European or Japanese societies that did this investing. It was the work of the industries and services that had originated there. It is useful to remember that, while the brief recession of 2003 hit the U.S., Europe, and Japan fairly hard, the (now) global industries that had once been called U.S. or European suffered very little.

Industries were becoming truly global organizations, yearly increasing the social distance between the corporation and the country where it had originated. Businesses were investing everywhere they saw the potential for market growth or lower production costs. In the free trade regimen of the time that meant that few products had clear national origins any longer. From consumer electronics to passenger airplanes, design, component manufacture, and assembly took place globally. For truly large multinational corporations, this globalization offered some remarkable advantages. The economies of scale possible with global markets kept costs down as did the flexibility to locate manufacturing in locations with low labor costs and investment incentives (like minimalist environmental regulations). Large global corporations were in a position to flexibly shift products into whatever markets were experiencing the most growth, and to shift production to mitigate global currency fluctuations. These trends led rather inevitably to global consolidations in most industry segments — bigger was definitely better.[2]

While many factors contributed to global growth, it was the nexus of telecommunications and air transport that made the world hum. Confounding the predictions of many computer industry pundits, the explosion in telecommunications actually ignited air transport into a global growth surge. Part of the reason, of course, was the somewhat chaotic state of the telecom-internet system. In the late 1990s and early 2000s it exhibited all the best and worst features of a new free market technology. New network performance was constantly being added while a plethora of business and personal applications flooded the marketplace. The pace was so fast that industry standards and protocols actually fell behind. This was compounded by the fact that the business

---

[1] Of course, we can now trace a few of the less effective terrorplots perpetrated on society in the coming years to the anger generated by these "minor and regrettable social service problems."
[2] The disaffection many in the world felt toward these huge corporate monoliths that cared only for profits in a shifting marketplace not only contributed to the early terrorplot target lists, but also slowed the public outcry in the very early days of the terror.

model that maintained cash flow into the network owners was increasingly under pressure. Competition from new entrants and new technologies plus price wars led to several bankruptcies and inevitable disruptions in service globally. The result was that businesses could make excellent use of the network, provided nothing crucial was committed to the ether. The need for personal contact grew exponentially just as a host of new markets and opportunities were emerging across the globe.

Passenger aircraft, cargo aircraft, specialty air vehicles — faster, bigger, more flexible, more reliable. The market demand was enormous and driven by the needs of international business and an explosion in leisure travel. Spreading global affluence gave many the ability to explore places they had only seen in film — the Castles on the Rhine, the Great Wall of China, and Graceland. It was a time of nearly unprecedented building of global infrastructure, especially in the emerging markets. New national carriers and aerospace firms emerged as demand drove the installation of new production capacity. They stood on their own[3]; they allied with the giants of the earlier days, and often they merged, as consolidations swept the aeronautics industries.

Of course, this kind of rapid growth left many people and groups behind (only temporarily, it was hoped). In the old industrial nations, the unskilled and semiskilled saw no great advantages to global economic growth. Indeed, the loss of jobs to the emerging markets had hit this group of poorly educated quite hard. They found only spotty relief in service industries that paid modestly well, but demanded long hours, some learning skills, and a very customer friendly persona. The rapid growth of consumerism was not to the liking of many highly conservative religious and cultural authorities, and an interconnected global economy robbed many authoritarian government leaders of the total sovereignty they thought was their due. Corruption, as well, was to be found in many of the most turbulent markets. These issues were not in the field of view of most consumers and voters, however. The world was getting better all the time, and all small "adjustments" would take care of themselves — as they should in a free market system. In a sense, they did take care of themselves.

Considering the dynamism of the age (or perhaps because of it), when the world changed and our lives were altered so profoundly, it happened surprisingly slowly. The change was so significant, that it seems somehow wrong that it should have appeared almost evolutionary. There should have been more high drama — an off-stage narrator who says (in deep and sonorous tones) " . . . and now the world has changed. . . " Yet, the truth was, it took years for it to sink in that our lives might well have been remade forever. My colleagues studying Crimes and Terror in the Department of History debate endlessly on "The Origin." I, for one, am content with the conventional wisdom that places the beginning of our era at the Ramadan Festival in Medina in 2007. Six thousand Muslim pilgrims dead from the release of mustard gas, while across the globe, over the next four hours, six passenger planes carrying Muslim pilgrims were bombed No one ever took "credit." At the time we thought that was unusual!

---

[3] Occasionally they had local government subsidy, but also that became increasingly rare as the global free market ethos took hold everywhere.

There was a hue and cry throughout the Islamic world and all caring people of the globe reacted with a mixture of horror and deep sympathy. The initial actions were classic and frightening at the time. Air passenger traffic fell off dramatically for a while, Israel was threatened with invasion, oil was briefly cut off to the West, several Western corporate buildings in the Middle East were destroyed in rioting, and calls for a jihad (but against whom?) were heard from Manila to Morocco. Yet, in the end, there was no evidence; no one to blame. To this day, nearly fifteen years later (and despite huge investments in research), no culprit has ever been brought to light. The most amazing theories continue to circulate about "the origin." It was the Masad; it was the Sword of Isalm, it was the neo-Nazis; it was the Yakuza; it was oil spot market traders. The most interesting thing about the "origin list" is that governments never seem to be seriously considered — and *that* perspective was to be a harbinger of things to come.

Despite grief, fear, and anxiety, the call of global commerce went on. Within a few months, the world was back to its normal hectic pace. A Federal Express plane bound for Memphis was destroyed in a bomb blast, but that received little attention despite the fact that, once again, no one took credit and no one was caught. Then heightened security at airports paid off in a big way. As the Christmas air passenger rush was just beginning, bombs were discovered simultaneously at Heathrow, Kennedy, Chicago, Miami, and Rome. Flights leaving Boston and Frankfurt were not so lucky and five hundred people, mostly families, died in explosions. A pall fell over the world and air traffic in particular as panicked holiday travelers called off their travel plans (especially those flying internationally) Once again — and not for lack of trying over several years — no one was ever caught. Two weeks later, despite very tight security, two flights out of Bombay were destroyed in bomb blasts fifteen minutes apart. This time the perpetrators were caught. Amazingly (at the time) they did not know each other and their efforts had been entirely independent. In one case, a Mr. Madan, had simply decided to use the growing terrorplot mania to cover up the murder of a troublesome business partner. In the other case, the target was the airline itself — destroyed by an ex-employee, fired for a history of "antisocial behavior."

Not surprisingly, Amtrak, Eurorail, and in fact all land-based mass transit began doing a land office business. Or they did until an Islamic fundamentalist group blew up a train inside the Channel Tunnel, while just a month later a survivalists' cult tried to blow up an Amtrak train because it was carrying "new Eastern settlers as agents of global capitalism." Nothing, it appeared, that carried a large group of people, was safe. And that fact — that places with large gatherings of people were not safe — was to become the driving force of our lives.

Stadiums and office buildings, train stations and outdoor concerts, movie theaters and shopping malls — the list of places attacked worldwide over the next five years was as varied as the forms of attack. While there were a few "lone gunmen" with automatic rifles, most forms of terrorplot attack were "impersonal instruments" of large-scale death. Bombs, missiles, and gases were the most common in the beginning, but, as security improved, a more insidious form of attack emerged in the use of airborne infectious diseases. These were especially effective on passenger planes — and airplanes (for all methods of attack) were always the favored target for the publicity (and the degree of difficulty in "succeeding").

Targets and approaches were not the only things that varied widely, so did the reasons for the attack and the successes at capturing the guilty. Although many were caught, it was very hard to find a guilty party when motives (and credit-taking) were either confused or absent. Imagine a passenger plane has crashed into the ocean. The last few transmissions from the pilot documented high fevers and dizziness among passengers and crew.[4] Provided you can find the wreckage, how do you find the culprit? No one admits responsibility. Was it a suicide (to provide insurance money to a family), was it a neo-Nazi effort to destabilize politics, was it a disgruntled ex-employee of the aircraft fuel delivery company, was it done just for the challenge of succeeding, was it done by a company who sells remote diagnostics equipment, was it done out of racial hatred, and on and on . . . As the number of unsolved terrorplots went up, the number of people and organizations attracted to exploiting the situation went up. The threat of capture and punishment was not going to be an effective deterrent, so protection — elaborate and very expensive protection measures — were established.[5]

It became an age-old game of "offense/defense" race. New technologies and operations to protect and secure tried to remain ahead of new terrorplot technologies and approaches to mass murder. Eventually, the defense began to win, because the frequently "independent and unorganized" terrorplots did not have the resources to match corporate and government programs to provide safety and security. However the costs have been very high indeed — in traveler peace-of-mind, time, and money.

No one readily gets on an airplane today; indeed, it has become the case that most people just don't consider flying in the first place. We have not had a successful terrorplot event in four years, but that is no longer the point. Certainly significant residual fear is still in many minds. Despite the fact that security measures seem to be working well, anxiety remains high. As a matter of course, all alternatives to mass travel are explored before a trip is reluctantly undertaken. The more fundamental reason why we have turned from air travel (when possible), however, is the cost in time and money. And the time it takes! Deliver your luggage to the airport one day before your trip (two days for international travel). Arrive at the airport three hours before departure. Wear no metal or plastics. Carry no electronics. Don't plan tight schedules for the other end of your trip because departure times are randomized in half-hour segments. Finally, however, despite the anxiety and the time, it is the increased costs that have become the primary deterrent to extensive use of air travel. Users pay for the enhanced security (even government-mandated security measures) and all the services and technologies that have

---

[4] Of course, most readers will recognize this particular event and will remember the quite profound effect it had on the traveler psyche.

[5] It has been popular in recent times to criticize the "free market" approach initially taken in security measures. Most people look upon the more recent and very intensive government attention to safety and security as the model that should have been followed in the very beginning. But, it must be remembered, that was not the public attitude of that era. Government intervention was very much out of favor then. And, to be perfectly frank, it is quite possible that far more innovative security operations and technologies came onto the market faster, as a result of market forces, than ever would have been generated in a government procurement system. Costs based on user fees has been expensive for transportation users, but I will argue that efficacy and lives saved has been adequate compensation.

sprung up to help ease the inconveniences. None of this is a social cost defrayed by the taxpayer. The costs are all directly supported by the traveler — and they are high, very high.

The government's role, of course, has evolved in two complementary directions. Within a year of the 2007 Christmas bombing attempts, as global commerce was beginning to show little recovery from the falloffs due to reduced air travel and transport, the U.S. government launch a significant new policy initiative quite at odds with its laissez faire economic and social policies. Until the terrorists were caught,[6] alternatives to the mass transportation of people had to be found. The natural choice was the Internet, but it was not a reliable network — certainly not on a global scale. The U.S. government embarked on a massive program composed of subsidies, tax incentives, regulations, and multi-lateral negotiations to establish a Global Infonet as a reliable backbone for global commerce. In the U.S. the network took the form of a publicly regulated utility with some publicly funded applications (education and public service notices, for example) but mostly privately sold applications for business, communication, and entertainment. The theory was twofold: (1) provide a standardized and superbly reliable backbone for personal, government, and industry use (to "reunite" the country), and (2) subsidize its early globalization as a competitive advantage for U.S. business (to "reunite" the world on terms favorable to the U.S.).

Within eighteen months, the U.S. standard Global Infonet (the G-Net) had become the worldwide standard. With more and more bandwidth and applications constantly added (some at U.S. taxpayer expense), the G-Net was becoming the backbone of global commerce and an acceptable (and very cost competitive) alternative to all but the most essential travel. The global economy started to rebound and U.S. corporations, with "first service" command of key business applications, were well positioned for leadership. The tax money had been well spent.

The second government initiative was to "provide for the public safety." While the primary philosophy of the government remained noninterventionist, in the area of safety and security nothing short of a massive industrial and defense policy was initiated. Infrastructure, technologies, products and services — if they could improve the safety and security of people while they traveled or met in large groups, then either financing or funding became available. If the solution was applied technology, the marketplace was given free reign. If long-term R&D activities are required, both government and entrepreneurial capital are available. While the government continues to prefer a free market approach, public funding was and is quickly allocated for programs that will provide enhanced security. The mission and budget battles that have erupted over safety missions are fast becoming legendary.

People have adjusted their lives relatively well by 2020. They stay at home more and entertainment tends to be small-group or family oriented. Interactive G-Net programs are quite the rage. Businesses, especially those with a significant global component (which were many) have had a more difficult time. Some have simply failed. The costs of being global were too high as were the costs of shifting to a more regional focus. Some retain their global standing and

---

[6] At that time, of course, everyone still believed that they were dealing (mostly) with classic (probably state-sponsored) terrorism and that the guilty would be caught eventually.

are trading across the globe but at much higher operating costs.[7] Many others still consider themselves global companies, but they have significantly regionalized their operations to reduce transportation needs. Since economies of scale are not always optimized, costs are occasionally higher. Nevertheless, we have a vigorous global economy and all are adjusting to the new realties of global life and commerce. It is simply the case that we hardly ever think about getting on a plane anymore.

Postscript: Just yesterday (as I was concluding this introductory essay) my nine year old was reading a novel set in the 1980s. She was puzzled by some of the things she was reading. "Hey Dad," she asked, "what's a carry-on?"

---

[7] Cargo transport cost much less than passenger operations, but cargo, too, was often targeted and has not escaped many of the security costs associated with all air transport.

Scenario Matrix - Grounded

Scenarios / Drivers	Grounded
World Economy and Market Environment	Personal security is paramount; global, relatively-free economy with some select government intervention to control terrorism, and coordinated, international, US government-led cooperation to build information highway; the information highway in many cases has significant cost and security advantages over traditional transportation; limits to the free market defined by threats to the movement of people and goods; US dominant economic power in part because of leadership in info highway.
International Trade Environment	International cooperation driven by security; high degree of harmonization on standards; the World Trade Organization works to restrain government involvement in nonsecurity trade and commercial matters; highly controlled access to airspace and landing rights; government run aviation security infrastructure, with some outsources to private technically oriented corporations; high degree of commercial harmonization for nonsecurity related activities; minimum standards on security must be followed by everyone.
Political Instability	Very significant terrorist threat; some threats acknowledged, some not; threats are highly random; reasons are varied - religion, economics, ethnicity, etc.; low to high technology threats (e.g. bombs, software viruses, bio terrorism); occasional collapse of authority in least developed countries; significant personal fear.
U. S. Military Requirements	Low level traditional military requirements; quick in-out police actions; significant excess global arms inventories, poorly accounted for; DOD assigned some non-traditional anti-terrorist roles; intelligence community well-funded; many small prepositioned stock-piles.
Global Distribution of Power & Technology	High tech, wired world - broad band everywhere and infrastructure development widely promoted - IT is the backbone of global economy; few barriers to technology proliferation; source of global power is economic and technology; large corporate security forces common.
Fuels & Fuel Sources	Oil available at moderate prices, price has gone up some over years; resources available and market driven; some short term variation in supplies due to related terrorist

Appendix D

Scenarios / Drivers	Grounded
	activities.
US Policy	Basic policy has shifted from laissez faire to more regulation and intervention; early government reduction giving way, increasing size in all areas that can claim a safety/security role; global trend to harmonize regulation surrounding corporate behavior (e.g. bankruptcy, anti-trust), still national variations exist; moderate liability; no tort reform; due to the government commitment to reduce the effects of terrorism there is major government investment in information infrastructure and reduction in traditional government technology investment (e.g., aviation); deficit reemerges (spend what is needed); low US unemployment; large multinational corporations have limited physical mobility and are vertically integrated; barriers to entry/exit generally low except in issues of security/safety then value hurdles are very high and often politically manipulated.
Corporate Structure and Operations	Global companies use the internet as substitute for travel, enabled by virtuality; limits to geographic consolidation, but economic consolidation continues; market share in products tend to be stable but market share for services on internet is dynamic, generally off-sets are reduced; strong global economy; alliances are important especially for market growth; ownership is global and dynamic; labor competition high at professional/managerial level but tighter at production level, since tied into local labor pool, with some what higher costs; hurdles in cost and location, but plenty of startup capital available.
Environment	Environmental issues at the global and national level are over shadowed by terrorism and political instability (but might be environmental terrorism); travel oriented pollution is down, but overall global pollution is up and is seen as a looming problem.
Public Health	Important if there is national security impact for example biological terrorism on planes, trains, boats, water supplies, etc.
Public Attitude to Technology	At the individual personal level there is a love/hate feeling toward technology, because technology is both the cause and solution of the problem; a possible technology proliferation race with terrorists that includes "information terrorism".
Education	Do not put all kids in one building; wired and virtual

Scenarios / Drivers	Grounded
	education has gone very far on global utility internet; socialization side of education is a problem; access to education is through public funding; higher education tends toward private funding; despite virtual education, degree granting still through organized "universities"; strong science and technology, arts and humanities tends to languish; the improved access to education through the utility infonet does not ensure that the underclass is well educated; US remains focus of world-class graduate education.
Geographic (Living) Dispersion	Security concerns define nature of dispersion (e.g., gated communities), dispersion distances determined by secure means of transit; some can telecommute all year.
Communications and Information Technology	Free-market development of telecom for first 10 years, then government steps in and establishes a high band width global utility network and standards; it becomes the heart of global business; crucial to mitigating the impacts of terrorism and high cost of business and personal travel; communication and information security, like physical security, is important.
Production Cost Performance	Strong aeronautics industry growth in 1st ten years, then excess manufacturing capacity in large commercial transport, leads to replacement and maintenance market; product improvement (and cost) is directed towards safety and security; operation patterns focus on safety and security; very high cost of travel becomes driving factor as terrorism is somewhat controlled; if firms own unique technology solution, some barriers to entry may come down; ownership pattern is private, vertical integration in many transportation and travel industries to control safety and security; there is a tendency toward somewhat synchronized global business cycles.
Technology development and Application	The need for security drives government industrial policy to augment market driven technologies in safety, security, and information/telecommunication; need new manufacturing technologies for cost effective small local factories; R&D is widely diffused geographically and across industries, "R" is more focused on safety, security, and communications.
Time Poverty Leisure Time, Entertainment	This is a moderate time poverty environment; security and transportation cost issues slow commerce down while

Scenarios / Drivers	Grounded
	security needs add time constraints; transit time needs to be highly productive; large group entertainment is not "physical" but virtual; personal entertainment activities are popular; security concerns reduce work environment mobility; networking encourages blurring of work, education, and leisure; significant home entertainment.
Global Transportation Infrastructure	Trend toward safe and secure transport infrastructure that supports "low density" vehicles; more difficult access to infrastructure to ensure security; infrastructure "guarantees" safety and security of transport; infrastructure must accommodate low-cost, rapid-time-to-market, optimized production flows; supports rapid response to consumer demand (internet orders/drop shipment delivery).
Safety and Security	Safety and security (including data security) is the essence of this world; it is on everyone's mind and drives cost up and convenience down.
Access to space	There is a practical attitude towards the use of space; projects that demonstrate practical applications can be financed.

Regional Tensions

U.S. Economic Competitiveness	Worldwide Demand for Aero Products/Services	Threats to Global Security and/or Quality of Life	Global Trend in Govt. Participation in Society
Strong	High Growth	High	High

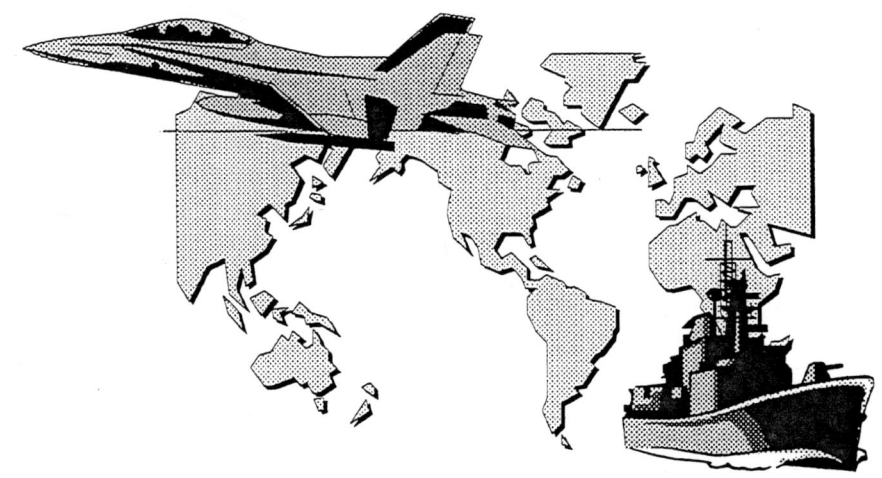

Summary

The world is organized along regional blocs each focused on military alliances but also containing formal and informal restrictions on commerce. China after a bloody civil war has emerged angry, aggressive, and particularly antagonistic toward the U.S. and all that it has stood for in the post-WWII era. In response an Anglo/Americas bloc has emerged which now interconnects North America and the UK along with Australia, New Zealand, and Taiwan. This bloc is politically leary of China but, other than Taiwan, is under no direct military threat. Japan, on account of growing doubts about U.S. trade and defense commitments, has moved closer to Russia and along with India has formed an alliance to contain China. Major U.S. companies, in the face of losses in China and hostility elsewhere in the region, have abandoned or downsized Asian operations. An expanded European Union (ex-the UK) is relatively strong, cohesive, and independent to the extent that it strives not to take sides and sells to everyone. Bureaucrats everywhere no longer apologize for interfering in markets when national defense is at stake.

The Futures Group

A century and a quarter ago, as the progress-obsessed nineteenth century was preparing for the *twentieth century*, the optimism was infectious. The scientific, economic, political and spiritual realms seemed to be harmoniously working together to advance knowledge, raise living standards, ensure peace, and spread "Western Civilization" worldwide. The new century not only promised technological marvels, expanding international trade, and increasingly liberal politics, but also higher culture and a more profound human spirit. It is only with hindsight that we can see the irony in those brave sermons and editorials that promised the twentieth century would see the material and spiritual ascent of mankind. *It is with similar irony* that we can now look back (from 2020) only twenty-five years at the descendants of those same prophets who were hailing "a new world order" and "the end of history." Their new world, we were assured, had learned from past bloodshed. Now "liberal capitalism," "global free trade," and a proliferation of "multilateral treaties and organizations" would ensure peace, progress, and prosperity for the new millennium.

Over the last quarter century, America easily had the world's most open, least regulated, and most globally integrated economy. With our deficit well in hand and our debt coming down, we were a nation filled with optimism about the growth of global trade and institutions. After the UN reforms of 1999, all past due contributions were paid and our foreign policy was increasingly coordinated with and through that organization's strengthened institutions. U.S. media giants distributed news, entertainment and educational information worldwide, with an explicit bias toward enhancing international understanding and tolerance. Our universities were open to the citizens of all nations, as were our information network, our financial system, and our corporate board rooms.

If the rest of the world seemed reluctant, at times, to follow America's renewed sense of humanitarian optimism, few countries were reluctant to engage in global economic growth, once again fueled by the North American economy. Trade increased as did all forms of social and economic contact. To the consternation of many, those contacts seemed to bring as much tension as international understanding; but that was easily explained by differential economic growth and the defensive cultural and ethnic policies pursued by some nations made uncomfortable with rapid globalization. All of that would pass (we thought) with the greater understanding that comes with time.

Many, of course, have pinpointed our nearly obsessive pursuit of economic globalization as the root cause of our inattention to politics — the raw sort of power politics — that we eschewed; yet was emerging again on the Asian mainland. The post-Deng succession *seemed* to go off without a hitch in the late 1990s. Or, in hindsight, perhaps that is what we wanted to believe. There *were* those rumors that spread through Hong Kong of assassinations and skirmishes between provincial armies. After several years in which Beijing seemed to be in control of a slowly (very slowly) liberalizing country that absorbed all the investment sent its way, China exploded into Civil War between 2006 and 2008.

In hindsight, the succession had not been all that smooth after all. Years of bitter power politics had remained hidden from the capitalists that scoured the country for investment opportunities. The sides of the struggle proclaimed differing ideologies — and perhaps they meant it as they sought control of the world's most populous nation. One side (the North and some of the interior) sought a return to communal values but that was mixed with fierce nationalism. The coastal provinces and parts of the southern interior sought to continue the Deng economic revolution and were considerably more internationalists.

A horrible war of mass armies using chemical and biological weapons was fought mostly in the interior and north. The "entrepreneurial" coastal provinces won with surprisingly little damage to the industrial economic infrastructure. But in a turn of events not unheard of in history, the winners began to mirror-image some of the loser's ideology. The emergent and reunited China faced the world as a ferociously nationalistic power hardened by war that immediately sought recognition of its (appropriate) dominant role in Asia.

It is difficult to blame the Europeans for being less idealistic than we. Twice in the last century they were torn to shreds by war. As many argue, however, there may have been more at play than fear for them to have shattered the most successful military alliance in history. Whatever it was, a lack of courage, jealousy, opportunism, or just a different sense of the politically possible, when NATO resolve and unity came to the test in the defense of Taiwan, Europe opted out. Only the UK — in part to finally protect its refugee Hong Kong constituency and in part to assert its independence of a Europe dominated by Germany — (and as always, Australia) chose to stand with North America against China's overt attack on international law. After three days of "live fire" intimidation, China backed down — as the Sixth and Seventh U.S. Fleets came within strike range — but its subsequent behavior toward neighbors (unwilling to resist) and toward foreign investors (all too willing to negotiate anything) has shown again that appeasement only perpetuates aggression.

Internationalism and, more importantly, multilateralism began to fail; incrementally at first, but fail none-the-less. As the joint command in NATO became increasingly fractious, the EU began shifting its focus to the European Defense Union to which North Americans were soon denied observer status. In the UN the excesses of the Sunni Theocracy in their new Arabian Republic were met with conflicting European responses. Suspicions and accusations that German and French oil companies had successfully negotiated secret supply contracts below the $40 marker price continued to surface. Perhaps more importantly, all attempts to embargo China and even curtail its trade violations through the WTO were met only with subterfuge and political sandbagging. Without multilateral cooperation, the US quickly discovered that its defense commitments were overextended.

At the same time, with the WTO hopelessly deadlocked, unlicensed and counterfeit goods began flooding the world market with no coordinated response. There were two major air crashes traced to "fully documented" defective counterfeit engine parts produced in China; and, yet, many governments clung to the belief that unilateral negotiation could protect their interests and investments. Left with no other recourse, America moved to impose maintenance quarantines in Asia and selective import restrictions.

These actions may have been poorly managed, but coming at the same time as our discussions with Japan about a reevaluation of existing defense commitments seems to have been the

Appendix D

deciding factor in bringing the New State Socialist Party to power in 2012. Their successful negotiation with Russia for a return of the Sakhalin Islands combined with significantly increased investment in joint development projects was the kernel of cooperation that became the foundation of their defense agreement against China. With India's increased participation in this alliance, they have become the dominant political and military power surrounding China and have successfully co-opted cooperation from some other Southeast Asian countries. The only exception being the Anglo/American presence in Taiwan, the Philippines, and, of course, Australia and New Zealand.

Thus, in less than a decade, the entire post-WWII international institutional structure is on the verge of total collapse. Just as the last semblance of the world created at the Congress of Vienna was totally destroyed by an economically strong, newly centralized and politically aggressive Germany in WWI, so too, a similarly strong, newly re-centralized and assertive China seems determined to do the same. Europe, however, is now the neutral, standing on the sidelines, trying to maintain relations with both sides; while the Anglo/American Alliance watches its Asian counterpart every bit as enigmatic and mysterious as its Czarist predecessor of a century ago. In a world with as little trust as ours, weapons, more than allies, provide security.

Unfortunately our resources are stretched thin: we have allowed our arsenal to lapse into disrepair; our economic muscle has been directed at more peaceful pursuits. Our failure to provide employment for a once unmatched defense industry has resulted in loss of many critical skill sets. We currently are deploying weapon systems that, today, we would find hard to produce. There are whole new areas of weapons technology, most notably in advanced information and knowledge systems, where we have trained the world but failed to retain sufficient talent to support the needs of both industry and defense.

In addition, the radical shifts in global economics and world trade have made all too obvious the folly of our excessive dependence on foreign production and our excessive reliance on overseas design and development. Luckily America's immigrant traditions are making it possible to aggressively "patriate" needed talent worldwide. Finally, despite the depressed condition of financial markets, the soundness of our recent fiscal policies have to some degree strengthened our borrowing power. At least there is now general consensus that society and the government have vital national interests concerning the nature and structure of the country's economy; laissez faire may be fine for angels but it is shear folly in a world of human conflict.

Despite this consensus there is no clear agreement regarding the nature of our defense requirements. Seemingly endless lists of needs and technologies are demanding our national attention: air and missile defense, rapid deployment forces, infocom-defense, space defense and global surveillance, chemical and biological systems, and on and on. In addition, the state of our intelligence capabilities has also fallen way behind what is needed. Luckily, the combined naval resources of the Anglo/American Alliance can provide the foundation for the kind of traditional sea-based power projection that appears to be needed. It is also rumored that the U.S. and UK intelligence operations are being successfully revived. The first real test of these joint capabilities appears to be coming in Africa where access to critical raw materials is once again under threat of revolutionary forces with foreign power backing — both the Chinese and European, if regional press reports are to be believed.

One of the biggest problems facing the Anglo/American Alliance has been developing a secure and integrated information network. The U.S. commitment to the Internet and the World Wide Web, like its other commitments to open and transparent international systems, left the country somewhat exposed. At the same time, the UK found itself tightly tied into the European IT System. Meeting all these new alliance-based objectives will provide sorely needed employment to counter some of the jobs that continue to be lost due to contracting world trade. In all this restructuring our only consolation appears to be the fact that the other world players are, in their own ways, equally far behind.

The Russo/Nippon aerospace consortium is still struggling to successfully integrate Japanese standards of manufacturing excellence with Russian design and material expertise. Intelligence reports continue to suggest that cross-cultural tensions are a major problem. No such problems exist in Europe, but there the focus is more tactical than strategic, and the goal is to simply be low-cost producer of technically acceptable armaments and transport equipment for regional and global markets. China's product is consistently better than the experts expect it to be, but the quality of its high-volume production goals are just as consistently less than demanded. Their reusable launch vehicle program is still suffering from the calamitous failure that killed twelve score or more of their top space scientists and engineers.

Despite all this global military buildup, many people are arguing that the greater threat may in the end turn out to be just as it was portrayed by H. G. Wells back when the twentieth century was still the promise of the future. Underfunding and deteriorating international relations combined with near-term "restructuring objectives" have led to neglect of most public health infrastructure and significant abandonment of environmental restrictions in much of the world. There are those in Japan who argue even more malicious motivations are at play and insist that the acid rain out of China is a premeditated environmental weapon aimed at destroying Japanese rice production. In a similar fashion Euro-hawks have argued that their people are most at risk to ozone depletion and potential climate change, but to date they have had no impact of EU trade policies.

The potential public health problems, however, may be of more immediate concern. People in the medical profession are quick to remind public officials that the swine flu epidemic of 1918-1919 was a far greater killer than all of the carnage of WWI, and up until the 1950s the biggest killer of all was TB. They then point out that reports began appearing in the early 1990s about drug resistant strains of TB and other diseases of urban congestion that were staging a slow comeback among the extremely poor throughout the world. Then in the first decade of this century an ever-increasing number of "exotic" tropical diseases began appearing in large cities of major industrialized nations; occurring, it was alleged, because of cheap air travel to and from emerging market countries. The reported use of biological weapons in the Chinese civil war may or may not have aggravated the already deteriorating situation.

Appendix D

Whatever the historical facts, most medical authorities are now arguing that of the Four Horsemen, plague is now a bigger threat than war and that far too large a portion of America's and the world's limited resources are being devoted to preparation for the wrong battle. Skeptics of course argue that such claims are little more than antiwar rhetoric in a new costume, just as environmentalism is no more than anti-industrialization. Much of the public, however, seems to be personally concerned.

The decline in long-distance vacation travel continues, and travel industry experts will tell you, confidentially, that it has to be more than the economy. They point out that luxury travel to exotic resorts is way down and even traditional Caribbean destinations are suffering, while U.S. and European destinations appear to be picking up. International business travel also continues to be soft, as corporations abandon any pretense at global organizational structures in favor of a more regional focus. But travel industry people say there is more going on: "Folks come in asking if the water is safe, if the rooms are clean or are just distrusting the 'people there.' And, of course, they invariably go on to add "concern about regional air safety and security management in one country or another." In the end they simply go some place local that they can trust. People, like their governments, do not seem to be in a trusting mood.

Whatever happens, the new millennium looks more like the "old world disorder" than anything else. And, if the statesmen or the public health officials can't keep the lid on things, the "end of history" may have a more final meaning than originally was intended.

Scenario Matrix - Regional Tensions

Scenarios / Drivers	Regional Tensions
World Economy and Market Environment	Slow growth in global economy with periodic instability; several large somewhat equal regional markets; multinationals forced to adopt regional structures; turbulent labor markets with overall unemployment decreasing from very high levels, moderate to high inflation resulting from declining trade and deficit spending; heavy trade barriers between regions; markets optimized for intra-regional relationships; have and have nots in all regions; little or no incentive for global harmonization; US lags in growth and competitiveness; globally, cost of capital high; little research & development except in military; risks are high.
International Trade Environment	Lethargic international trade environment; World Trade Organization is gone; trade agreements few and bi-regional; lots of trade barriers (tariff and non-tariff); tight controls on regional military-critical technologies; highly politicized air rights agreements; no incentive to continue off sets.
Political Instability	Four antagonistic regional powers - North America, Europe, Japan/Russia/India, and China that emerged from prolonged period of economic, political and military conflict; significant instability lingers in Southeast Asia; shared tensions over resources and concerns over sea lane security; China and US narrowly averted military conflict and are still at odds.
U. S. Military Requirements	Need for long-range force projection but no reliable overseas basing; Navy plays major role; growing DOD budget with classic missions, combined with new military needs including space defense and surveillance; no war at moment despite high tensions.
Global Distribution of Power & Technology	Regionalized; no dominant super-region; power fairly equally distributed among and within regions; not a globally wired-world; nets are regional; US retains technological edge.

Appendix D

Scenarios / Drivers	Regional Tensions
Fuels & Fuel Sources	Conservative Sunni Theocracy takes over House of Saud, Middle East oil goes to $40 per barrel (in today's dollars) but supplies are stable; balance of regional power prevents any reaction to price hike from consuming nations; high incentives to achieve energy independence for national security.
US Policy	Relatively high degree of government intervention in economic sphere; particularly strong industrial policy in defense matters; somewhat more liberal tort environment but not central issue; unemployment mitigated by rising defense expenditures; fairly permissive anti-trust regulations, activist government support for troubled US and North American companies; very high barriers to entry/exit ; entry barriers related to high tariffs, non-tariff barriers, and high cost of capital; high government spending, focused on defense sector; government debt and deficit both rising following relatively austere fiscal period; increase in regulations related to regional/international trade and commerce; higher humanitarian assistance which is pragmatic and directed toward Latin America to woo into North American sphere of influence.
Corporate Structure and Operations	Corporate structure deglobalized and now regionalized; high geographical consolidation with lots of corporate failures; very limited technological proliferation, very strong government controls on military and technology exports, but some seepage nonetheless; offset agreements the exception and determined by type of product and host country relationship to seller; business alliances are limited overall but critical to inter-regional commerce; pronounced trend to regional ownership structures; shortages of skilled labor especially for defense-related industries; technical labor costs very high, intensive training required to upgrade skill base; government support for defense related training and education.
Environment	High pollution globally, some regions better than others; not generally seen as social priority; pollution as "weapon"; pollution regulations as a

Scenarios Drivers	Regional Tensions
	non-tariff barrier.
Public Health	Public health and potential public health problems are consistently worrisome across globe; unified public health standards in Europe, uneven elsewhere; communicable diseases successfully managed at regional level; crumbling US and European sewer and water infrastructure heightening health risks.
Public Attitude to Technology	Attitude is positive and re-enforced by government propaganda; "Superior" technology is developed in North America; national security-related technology race between regions.
Education	Public funding based on national security needs; funding high but competes with other social requirements; targeted spending on science and technology; for universities, far fewer foreign students (situation very tough on private colleges); US higher education not necessarily the best and funding is constant problem, private university closures; academia might be key forum for global conferences; little technological innovation, stuck at 2005 level; classic educational paradigm; global harmonization of education in decline.
Geographic (Living) Dispersion	As offshore manufacturing returns home, government provides incentives to revitalize economic hardship zones; knowledge workers able to telecommute.
Communications and Information Technology	Communication tends to be regionally organized; global communications has extensive barriers; technology development is often a spin-off of military needs; global system in place in 2005 is still there (maybe not well maintained) and there may be private nets developing; slow divergence of standards and protocols between regions.
Production Cost Performance	Economies of scale limited by regional structures, making regional-based manufacturing more expensive; many raw materials and components outsourced out of US in earlier years are difficult and costly to obtain; some cost issues tolerated due to high defense need; barriers to entry/exit are moderate to high; governments are biggest buyers of

Appendix D

Scenarios / Drivers	Regional Tensions
	aeronautics products and services; subsidies to airlines for Civil Reserve Air Fleet; governments are committed to keeping airlines in business; very high emphasis on dual-use components and products; global business cycles not synchronized.
Technology development and Application	Government industrial policy focused on national security applications; core competencies vary across regions and focus on legacy industry base and local raw materials (new technologies developed to make best of local materials); government R&D is everywhere - it is both R and D.
Time Poverty / Leisure Time, Entertainment	Some time poverty for people working two jobs in lethargic economies (different from region-to-region); pronounced local and regional orientation to entertainment and leisure activities; classic pattern of work and leisure; some tendency to use entertainment for propaganda activity; global competition played out in art and entertainment fields.
Global Transportation Infrastructure	Infrastructure supports national security mandates; regional infrastructure predominates while inter-regional standards diverge; some deterioration owing to lack of maintenance of infrastructure not critical to national security.
Safety and Security	Security that is important is national security; safety is regional responsibility; harmonization in global standards are declining.
Access to space	Leading-edge space capability is central to national security concerns; space assets to support economic competitiveness protected by regional space defense forces.

Trading Places

U.S. Economic Competitiveness	Worldwide Demand for Aero Products/Services	Threats to Global Security and/or Quality of Life	Global Trend in Govt. Participation in Society
Weak	High Growth	Low	Low

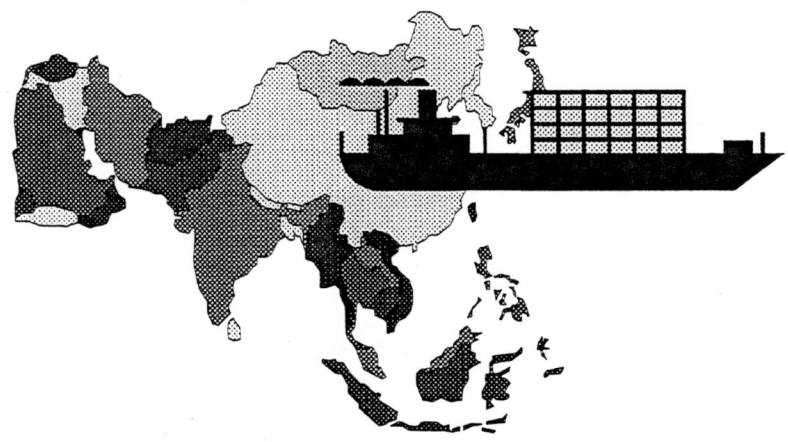

Summary

The past two decades have witnessed extraordinary shifts in global economic and political power. The booming economies of East and Southeast Asia have grown to the point of challenging the economic dominance of the U.S., Western Europe, and Japan. Powerful multinational corporations combined with free and open markets guided the flow of manufacturing, technology, and intellectual capital from West to East. Technological innovation, too, is increasingly centered in the Pacific Rim, as companies gear their research and product development to satisfy the wants and desires of the middle classes of China, India, Korea, Southeast Asia and, to a lesser extent, Latin America. The U.S. and Western Europe muddle through with sluggish growth, and heavy social burdens related to aging populations. Governments are pressured to do more — impossibly — with less. Privatization and deregulation offer only partial relief to strained government budgets. In the U.S. and Western Europe, unemployment is relatively low but underemployment is high. Consumers work hard and long hours, with little leisure time. For most, life is hard, but still bearable. There is a sense of inevitability to the tremendous changes that have taken place in the global economy. However grudgingly, consumers accept their diminished status. Mature market consumers are still the most materially fortunate in the history of the world. It is just that the rest of the world is now catching up at breakneck speed. Sooner or later, the "emerging" emerge; the "declining" decline.

The Futures Group

In 1996, *emerging* economies were all the rage. The original "Group of Seven" industrialized countries (which at that time included the United States, the United Kingdom, France, Germany, Italy, Spain, and Japan) saw great promise in the then-underdeveloped but rapidly growing economies of China, India, Indonesia, and the other export-oriented economies of Southeast Asia as well as those of Latin America. What the mature economies failed to realize at the time was the extent to which their own economic problems and declining competitiveness, in the context of free and open global markets, would accelerate the progress and development of the Emerging Markets and significantly narrow the gaps between the "haves" of the mature markets and the "not yet haves" of the Emerging Markets. The ascent of the latter has been no less than remarkable, as symbolized by China and Korea's admittance to the world economic power club. Today, it is known as the Group of Eleven (and growing).

U.S. and European investment in the Pacific Rim progressed at a steady pace through the end of the 1990s and the early 2000s. Though the majority of dollars, pounds, and marks found their way to Asia, select Latin American countries, notably Chile, received attention as well. Chile itself grew increasingly integrated with the Pacific economy. In the process, truly global companies emerged, with the ability to rationalize R&D, sourcing, production, distribution and servicing on a worldwide basis. The pronounced concentration of market growth in Asia Pacific resulted in a very large share of financial, infrastructural, industrial, and intellectual investment in that region.

Increasingly, the national origins of most of the world's large companies became difficult to determine. Trade barriers dropped — there was no point anymore in trying to shut out another nation's goods and services; there was room for everyone — as the explosion in privately developed information networks and widespread commercialization of the Internet greatly facilitated global commerce. The nascent but highly effective World Trade Organization kept all the players honest.

Across Asia and Latin America, rapid and sustained economic growth created a shortage of managerial and technical human resources. In response, both governments and multinational companies expanded sponsorship of advanced studies in the leading U.S. and European universities. Consequently, foreign (and especially Asian) enrollment in U.S. engineering and management science programs skyrocketed. So great, in fact, was demand for advanced training that many top flight schools opened up campuses in Asia. It was a big deal when Harvard announced the opening of its Shanghai satellite campus (the "Harvard on the Yangtze"), but by the time Yale, Princeton, and UCLA followed with their campuses in Santiago, Kuala Lumpur, and Jakarta, respectively, few seemed to notice or care.

In the West, the U.S. and European economies appeared deceptively healthy at the dawn of the twenty-first century. To be sure, both regions were benefiting from the robust trade with the developing world as lower labor and manufacturing costs that came with the move of factories offshore caused corporate profits to hit record levels. U.S. and European stock markets soared. In Washington, tight fiscal policies enabled the government to stem the increase in the national debt. To many citizens' amazement, the budget was actually balanced by 2002, as promised back in 1996, and the nation had begun to dig itself out from under its heavy debt burden. Confident in this perception of economic and corporate health, some leading-edge baby boomers started retiring at age 55, and didn't seem to mind when Washington made cuts in Social Security and Medicare benefits to help balance the budget. They didn't need to worry; their investments and retirement provisions promised a very comfortable lifestyle.

In reality, however, serious economic problems loomed just beneath the surface. By 2004, a large and rapidly growing share of the intellectual and physical assets of key sectors such as automotives and electronics was now located in China, Southeast Asia, and Latin America. The U.S. aerospace industry was hanging by a thread. Much reduced were the high-wage jobs that were the hallmark of these industries. This sparked protectionist opposition in the U.S., but no large-scale revolt. People still had jobs and managed to pay their bills. Besides, consumers had grown smart about world trade matters. They liked the inexpensive and increasingly world quality imported goods to which they long enjoyed access. The last thing they wanted was a trade war (which by the way would inevitably victimize some family member employed in an export-oriented industry).

True U.S. leadership remained concentrated in — non-manufacturing industries like information technology, health care, and pharmaceuticals. Though the U.S. suffered only one actual recession during this time (1997-1999), overall U.S. GDP growth averaged a paltry 1.5 percent between 2000 and 2005.

China and Southeast Asia, meanwhile, capitalized on the problems of the increasingly aging and noncompetitive markets of the West. Singapore emerged as a strong financial market after a series of trading scandals in Tokyo shook investment markets in the early 2000s, setting the government on a strict financial markets reform binge. Asian companies succeeded at gradually increasing their ownership shares in formerly U.S. and European manufacturing assets. Likewise, Asian companies and governments now accounted for a very high share of Internet traffic and, before long, took the lead in creating their own Asian language information networks and communication protocols.

The internal markets of Asian countries "took off." By the turn of the century, China boasted the world's third-largest market after the United States and Japan. Many Western pundits had an exaggerated sense of China's frailties.

Deng Xiaoping's passing in 1998 failed to create the tumultuous succession struggle in Beijing that the world had for so long anticipated, adding to China's ability to steadily grow its economy through the turn of the century. (At the fifteenth annual Brookings Institution *China After Deng* conference in 1997, leading Chinese scholars again predicted a prolonged succession struggle.) However, when Deng's successor Jiang Zemin perished unexpectedly in 2004 (when his Chinese-manufactured Boeing 737-1200 crashed off the coast of Hong Kong), neither Beijing nor the world was prepared for the brief but intense power contest that ensued. Under Jiang's rule, China's liberal faction had reemerged as a potent political force and was poised to make a play for party and government leadership. The military, long pacified by Jiang's modernization policies, grew nervous over the prospects of a reformist government. The military staged a feigned move against Taiwan, not so much to achieve independence (the Taiwanese and Chinese economies were so closely integrated by this time that even the People's Liberation Army recognized the dire economic consequences that would follow military action against Formosa), but more to reassert its influence in domestic Chinese politics.

Sadly, the United States grossly misinterpreted the Chinese Army's actions. Because of growing animosity over China's growing economic prowess fueled by the protectionist camp (as well as a massive outflow of talented China analysts from the U.S. Intelligence Community), the prevailing view inside the Beltway was that an invasion of Taiwan was imminent, and the U.S. administration was pressured to slap heavy economic sanctions on Beijing. The U.S. failed to marshal international support for its actions (European governments had correctly interpreted China's military maneuvers as nothing more than Chinese political infighting), and the economic sanctions had a negligible effect on China's economy. By the time the U.S. lifted the sanctions in 2006, its reputation as a reliable trading partner was greatly tarnished not only in China but also throughout Asia. A market of more than 2 billion people was virtually lost.

The convergence of a loss of Western industrial leadership in key sectors, growing Asian economic and political power, the trade dispute with China and its aftermath, and structural economic problems in the U.S. set the U.S. economy on a downward slope in 2006. With the bulk of the U.S. labor force increasingly finding employment in low-wage service jobs (want ads for the Gap appear weekly in the Sunday *New York Times* where solicitations for computer programmers and electrical engineers once ran), U.S. tax receipts declined. This coincides with the bulk of the baby-boom generation's entry into old age, placing enormous pressures on the tinkered-with-but-unreformed Social Security and Medicare systems. The increasing political power of this demographic group — the ranks of the AARP swelled by 20 percent between 2006 and 2012 — made it almost impossible to achieve structural reform in the entitlements system, causing the U.S. budget deficit and debt to balloon to levels that surpassed the debt of the early 1990s, sending the nation into recession. Unemployment approached 10 percent, underemployment was severe, and overall consumer confidence was at a 20-year low. In a much ballyhooed event, U.S. and European governments held a conference in 2011 to discuss ways to achieve drastic reform of their beleaguered social welfare systems.

Significant research and development capacity followed U.S. and European manufacturing overseas, and what little R&D was left was cut to the bone by short-term-minded corporate leaders trying to squeeze even more profits. Real economic recovery would not occur for another six years. At the same time, the foreign students that streamed into U.S. colleges and universities are now returning home to capitalize on the opportunities there at the same time that U.S. applications drop, leaving classrooms and laboratories — and endowments — virtually empty. College has become an increasingly unattractive option for many; the lack of high-paying jobs hardly makes the average annual tuition of $100,000 worth it. Very little new intellectual capital was being generated in the United States.

Meanwhile, Asian economies in the late 2000s truly came into their own. Japan got its financial house in order and was once again a strong investor in the region. China and Southeast Asia are on their way to becoming industrialized societies and have assumed solid leadership in several sectors, especially those that are technology and R&D-intensive. When the Shanghai-based

Xingou Technologies began mass producing and exporting its "biochip"— a device that transformed artificial intelligence into the real thing — China was finally able to tout its status as a world leader in both biotechnology and computer hardware. Before long, almost every Asian economy found its industrial niche: in Malaysia, it was automobiles, India, software, Indonesia, aerospace. In Latin America, Mexico, and Brazil were the regional hubs for automotive manufacturing; Chile was the leading developer of Spanish-language software.

The late 2000s also saw a dramatic shift in political and military power eastward. Distrust of U.S. motives in the Pacific Rim remained strong in the aftermath of the ill-fated U.S. trade war with Beijing, and few tears were shed when Washington announced that it would relocate more than 80 percent of its military assets in Asia back to the U.S. due to budgetary constraints. China, which had invested heavily in its military and now possessed state-of-the-art fighter aircraft and naval platforms, quickly stepped in and became the dominant regional military power. It resurrected the Southeast Asian Treaty Organization and goaded (somewhat forcibly) its southern neighbors into a rigid security pact; even Vietnam felt compelled to join. In a quirk of history, the new SEATO charter was signed on the very same day that NATO, whose members were unable to sustain their respective security commitments, announced the dissolution of its military pact. To the north, Japan, still wary of Chinese military intentions, struck a tenuous security alliance with Russia and India.

Fortunately, global security threats were relatively low. Ideologies have been replaced by the desire to cash in, and no major regional tensions are apparent. Even the Middle East was uncharacteristically stable after mild turmoil in the late 1990s, allowing for healthy petroleum supplies at prices that rise only modestly, in pace with inflation. The most significant military threat for the new Asian security system is the occasional flare-up of domestic turmoil in countries that have been left behind the economic wave. China has on occasion been forced to send peacekeepers to "the three B's" (Bangladesh, Burma, and Bhutan) when civil unrest threatened to interfere in regional trade flows.

By 2015, the transformation was complete. The U.S. and European economies had rebounded from their late 2000s downturns and, though hardly robust, were getting by. Consumers, however, exhibited a weary resignation to the extraordinary changes that had taken place in the global economy, and a sense of missed opportunities was pervasive. The decline in wages had leveled off, but the absence of well-paying jobs meant that most middle-class wage-earners were holding down two jobs, creating intense time poverty pressures. Job security was tenuous.

The U.S. technical and research infrastructure is tiny compared to what existed in the 1990s. A record number of private U.S. universities closed between 2011 and 2015, and the small percentage of middle-class high school seniors opting for college tend to favor community colleges over larger, four-year institutions. Corporate investment in R&D tends to be overseas. U.S. servicemen and women have been introduced to no new military platforms in nearly 10 years, and the services are struggling to extend the life of systems well beyond their intended longevity. The nation's infrastructure is crumbling as federal and local agencies are able to afford only patchwork repairs to the road, rail, and telecommunications networks. The only bright spot: the emergence of the U.S. as the world's most popular tourist destination for Asian vacationers. The best roads in the country can now be found in the greater Disney World vicinity, and the Orlando airport has displaced New York's JFK as the initial U.S. point of entry for the majority of international travelers.

Asian governments, meanwhile, are now moving quickly to the center of global political, economic, and military power and influence. Their economies are developing, their militaries modern, and they hold considerable sway over the rules of international commerce. Asian needs and interests are now figured into international aviation treaties and landing rights agreements. The International Civil Aviation Organization has moved its world headquarters to Tokyo.

The emerging have emerged; the declining declined.

Scenario Matrix - Trading Places

Scenarios Drivers	Trading Places
World Economy and Market Environment	Pacific Rim centric world economy; strong market forces; moderately high growth fueled by China and India; industrialized world experiencing stagnant economic growth; unemployment is low in emerging countries, moderate in US, high in Europe; significant under-employment in industrialized countries; low global inflation with efficient capital markets; reasonable cost of capital.
International Trade Environment	Globally strong and open trade environment behind effective World Trade Organization; US and European leaders fend off protectionist pressures; global harmonization is relatively strong; Asian leaders set rules on air and landing rights; low barriers overall but Western economies starting to demand offsets.
Political Instability	Generally stable and peaceful world but lingering after-effects of lost US trade war with China; world is organized around world trade and China & Southeast Asia at the core of global politics and economic power; some civil instability within countries not participating in global economy or with unreformed domestic economies.
U. S. Military Requirements	Global military budget declining; most funding is maintenance, repair, and service life extension program; little R&D; Russia-Japan and China protect major sea lanes; US has minimal presence; some co-patrols with Japan.
Global Distribution of Power & Technology	The heart of power and technology is in the Pacific Rim; multinational corporations have shifted design and manufacturing to local markets around the world; fairly high tech and wired world based on private communications infrastructure driven by multinational corporations.
Fuels & Fuel Sources	Peaceful coexistence in Middle East allows for stable oil supplies and free movement of petroleum; moderate price increases in line with inflation.
US Policy	Government promotes deregulation and privatization, but lower wage work and less job security result; government permissive in anti-trust cases; laisse-faire government position on bankruptcy; low entry/exit barriers; government is downsized and stable; fiscal policy is disciplined and

Scenarios / Drivers	Trading Places
	austere, some debt reduction accomplished; reasonably stable but sluggish economy; reduction in humanitarian assistance given low military commitments and austere government spending.
Corporate Structure and Operations	Globally integrated and networked firms of blurred national origin; corporate consolidation mitigated by low trade barriers, high competition, rapid technical proliferation, especially to emerging markets; free markets generally, but any off-set requirements likely to be ours; ownership complex - multinational capital with lots of portfolio investment; wide range of alliances among multinationals and national companies; in US, high tech labor pool reduced because of tendency of US-educated foreign nationals to return to their homeland; low skilled labor in ample supply, with sluggish wage gains; US students postpone higher education because of high cost, personal debt and uncertain prospects; weak US labor movement.
Environment	In US and Europe pollution is less of a problem since economic activity is down; emerging markets pollution is high and a growing global problem but not yet an issue for them.
Public Health	US public health spending static though global spending is rising as emerging countries aim to achieve Western health standards; problems with Ebola-type new diseases as emerging markets develop.
Public Attitude to Technology	Emerging markets- very pro technology "technology for technology sake"; in the US, public attitude is positive (but does not encourage public investment in R&D).
Education	US public education is as lethargic as economy and society; US graduate education still world class, but everyone sees it as declining; major US universities leverage strong reputations and globalize (e.g. place campuses in Southeast Asia); education in Southeast Asia is well funded and is a technocratic meritocracy; in US, middle class is primarily educated in community colleges and state universities; US students pursuing medical, finance, and law curricula, plus retail and trade industries; in emerging countries, applied sciences and management strong.
Geographic (Living) Dispersion	In emerging markets, some dispersion away from old industrial areas as firms seek lower cost labor and more

Scenarios Drivers	Trading Places
	modern infrastructure; in US and Europe, high skilled professionals enjoy high degree of geographic mobility owing to high world demand for specialized skills.
Communications and Information Technology	US no longer shaping the global information - communication system; world is well wired but US no longer dominates; applications and hardware tend to be made and installed in emerging markets first; superb equipment can be purchased from emerging markets vendors; bandwidth is adequate and keeps pace with demand; some Asian-led government regulation of communication standards; large multinationals set up private networks; significantly satellite-based and mobile.
Production Cost Performance	Manufacturing costs are low with global economies of scale but manufacturing is typically out of US; very hot market with general expectations with constant stream of new products (especially in emerging markets); product definition is tuned to non - US markets; in emerging markets barriers to entry/exit are low to moderate (some industry policy even in laissez fare environment); in US, barriers are restrained but where they exist they are complex and occasionally contradictory; barriers tend to be market-specific depending on priority of industry; moderately synchronized and slightly longer business cycles driven by emerging markets.
Technology development and Application	Very market driven; market need is defined by Pacific Rim consumers; R&D is out of US, tend to focus on "D" except in a few research-intensive industries such as biotechnology, pharmaceuticals, and agriculture; US excellence also in finance, banking, medical, and other service sectors; corporations may own great technologies, but it is developed and utilized out of US.
Time Poverty **Leisure Time, Entertainment**	Emerging markets are very hot markets with high time poverty; US and Europe are destinations for global tourism; in US, time poverty chronic and comes from multi-job families and unstable work situations; emerging markets' peak earners are workaholics, but next generation taking more frequent and more distant international vacations; even if small percent are leisure traveling, base population is huge.

Appendix D

Scenarios / Drivers	Trading Places
Global Transportation Infrastructure	Significant infrastructure development in emerging markets; standards are being set by emerging markets but may be some conflict / competition between nations; trend toward "smart" infrastructure; US infrastructure aging and deteriorating (except around major tourist and retirement locations); trend toward some privatization; moderate growth in business-to-business electronic distribution, slower growth on retail side.
Safety and Security	Information security predominates; personal and industrial threats limited; security focused globally on information and intellectual property.
Access to space	Space programs driven by pragmatic economic needs, particularly of emerging markets; financing and technology would be made available by emerging and developed markets alike.

Environmentally Challenged

U.S. Economic Competitiveness	Worldwide Demand for Aero Products/Services	Threats to Global Security and/or Quality of Life	Global Trend in Govt. Participation in Society
Weak	Low Growth	High	High

Summary

After years of negligence and abuse the global ecology has turned against the industrialized world. Economically harsh measures are required if the now obviously destructive levels of CO_2 emissions are to be reduced and global catastrophe avoided. Of all the world's major economic powers, the U.S. is among the least prepared — politically and economically— to make the needed sacrifices and is suffering accordingly. All hydrocarbon intensive industries are under merciless pressures: legal, financial, and social. Noncooperating countries, and there are a number of them among the less developed nations of the world, are risking military as well as diplomatic and economic sanctions if they do not similarly curtail their CO_2 emissions. The World Environment Council has become a dominant force in world affairs surpassing in importance and power even the IMF and the WTO. Despite the considerable expansion of governmental and regulatory intervention to reduce the CO_2 threat, public confidence is low and economic expectations are lower.

The Futures Group

Appendix D

The following excerpts are reprinted with permission of the Children's Education Project of the World Environmental Council, Brussels, 2021

. . . While all those requests have been more than flattering, the real reason I accepted this assignment sprung from an experience I had in my children's school last year. I deeply appreciate the confidence placed in me by the Brussels Commission, but the truth is I never would have undertaken this essay were it not for the discovery on my part that children take so much of today's life for granted, that they truly do not know how close we came (and still are!) to the extinction of our species. It is a story worth knowing.

The ethics of our era are unashamedly global; but I will be forgiven, I know, if I begin this story from the perspective of the United States. After all, this essay is first intended for that school audience; and Americans are clearly the ones who need the most help in adjusting.

To the surprise of all economic pundits of the 1990s, the U.S. gradually lost its global competitive position in business, technology, and sheer societal energy as the early years of the millennium passed. While many nations, especially in Europe and in the emerging markets, surged in growth, the U.S. economy stagnated as U.S. corporations wasted considerable investment capital and incurred huge debts in a takeover and merger mania. Wall Street's pressure for short-term profits resulted in significantly reduced funding for research and development across many industry segments. Moreover, increasing numbers of firms moved sourcing and manufacturing operations offshore to reduce costs and more effectively compete in the face of intense global competition. Many firms also were under pressure to maximize cash inflow to cover very heavy debt service obligations.

Poor business decision making was not the only reason for a lethargic U.S. economy. After a brief and unsuccessful attempt at fiscal reform, the U.S. government (under pressure from an aging population) returned to high social spending mixed with regulatory invention and increasing tax burdens. Of course, the U.S. government was unable to reduce the debt under these circumstances. In fact, the U.S. debt increased as various administrations were unable to control escalating Social Security costs resulting from intense political heat applied by a powerful and extremely well organized aging population. Healthcare reform efforts suffered similar resistance. Schemes such as federal requirements for U.S. corporations to assume a greater share of healthcare costs further encouraged corporate flight from the country.

Increased taxes for social programs resulted in low personal savings rates and less capital for infrastructure improvements and research. Efforts to reform education in the U.S. became embroiled in ideological disputes. As a consequence of offshore sourcing and inadequate training of the work force, unemployment among lower-skilled workers grew rapidly. Most new job generation was in the low end of the service sector.

The only real improvement in the U.S. economy was in trade policy — it came as a Pyrrhic victory, however. The U.S. made some early progress in reducing its trade deficit but it had done so using strong-armed trade threats and nontariff instruments, particularly in dealings with Japan and China. It worked for a few short years (into the early 2000s) while the U.S. economy was still the most attractive consumer market. However, as the consumption capacity of the U.S. economy waned (causing import demand to drop), the Japanese and Chinese retaliated with trade restrictions on of U.S. products. Antagonisms ran high, but the U.S. was powerless to do much more than live with reverse trade discrimination.

There was limited public support for defense spending since there were few perceived threats other than economic disputes with Japan and China and some intractable terrorist issues, particularly Islamic fundamentalism. Several U.S. failures in supporting nation building experiments in Africa and Latin America fed a growing national mood of isolationism. The resulting reduced defense budgets led to a decline in DOD procurements. Limited orders caused a major decline in the defense industrial base; suppliers to the prime contractors were particularly injured.

Other major economic powers did not share the U.S.'s problems. During this same period, Europe, Japan, China, and a number of countries on the Asian littoral experienced significant growth as a result of the improving efficiency of global capital markets targeted at investments in new and emerging global markets. As East Europe recovered from its depression in the 1980s and 1990s, it provided a major market for exports from the European Union (EU). Similarly, the growth of the middle class in such countries as India and China not only provided large internal markets but also provided major opportunities for Japan and other Pacific Rim exporters. In China alone, annual export earnings and GDP growth averaged 10 and 12 percent, respectively. Meanwhile, healthy economic growth permitted EU nations and Japan to make substantial progress in retiring the large social debt accumulated in the 1980s and 1990s. In Japan, for example, large corporate profits from increased exports permitted greater privatization of social services. In the EU, investment capital was available for infrastructure improvement. Japan, on the other hand, was able to divert money away from infrastructure projects having made heavy investment in the 1980s and 1990s. Japanese companies were therefore in a stronger position to take on long-term research and development projects.

A perceived reluctance of the United States to perform its traditional role in world affairs resulted in expanded defense spending in Europe and Asia. Although many Pacific Rim states had good economic relations with China, they continued to fear Chinese expansionism. China avoided a succession crisis and continued along a path of heavy defense spending (both in absolute and relative terms). Chinese military equipment modernization was unmatched by any country during this period. Japan, in particular, found the U.S. withdrawal from Asia disconcerting as it increased the vulnerability of trade routes and oil supplies, and generally disturbed the balance of power in the region. Japan amended its constitution to permit up to 5 percent of its GNP for defense spending and invested heavily in advanced naval and long-range air surveillance and interdiction technologies and capabilities.

Other Asian nations similarly decided that moderate to high defense budgets, particularly for acquisition, were prudent given Chinese capabilities. The EU, similarly, was disturbed by the seeming U.S. withdrawal from its traditional role in Europe. The potential for instability worldwide, possible threats to oil routes and oil reserves, and a requirement for a strong regional economic union motivated EU members to maintain a strong defense industrial base and dual-technology infrastructure. Europe increased its research and development budget. This effort to maintain a strong defense and advanced dual-technology base was partially funded through foreign sales, particularly to the Middle East and to emerging industrial powers in Southeast Asia. Many of these sales were through licensed production and technology transfer instruments.

Accelerated industrialization in China, India, and Southeast Asia was among the most significant developments of the late 1990s and early 2000s. Manufacturing grew at unprecedented levels; many of the Asian Tigers' manufacturing sectors grew at levels above 10 percent a year. It required extensive infrastructure construction. But the new availability of highways, for example, resulted in major demand for automobiles, in turn increasing congestion and pollution. Similarly, economic growth among these nations had the effect of creating a large middle class. The middle class became the major impetus for a large demand for all forms of transportation, including business and recreational air travel.

Russia did not experience the rapid growth of most of the industrialized or industrializing nations in Europe and Asia. Its government debt continued to grow, but it did make steady progress in resolving its massive economic problems and ameliorating social conflicts. Major European and Japanese investment in the manufacturing and resource extracting sectors was the primary reason for the steady improvement in the economy.

Growing industrialization worldwide generated enormous demand for Middle East oil. Global supplies remained adequate, however, and the major producers were unable to reach agreements that would have permitted them to significantly increase the price of oil. The main causes of this were the militant policies of Iran as well as major arms procurement by that country. All other nations in the region were compelled to invest heavily in deterrent systems. Thus the Middle East arms market was a major customer for the European defense industries. Similarly, China found a major market for its defense equipment in exports to Iran. Iran itself remained a fundamentalist Islamic theocracy. This kind of militancy did not spread to other Middle East states, however.

While global tensions were high, global business was expanding at a healthy clip, as evidenced by new factories, new trade and commercial relationships, and increased cross-border capital flows. This free-trade environment drove demand for truly efficient and global transportation systems. Predictably, living standards in the developing world improved as household income grew and physical and social infrastructures were developed. Diet and public health across the globe were moving toward Western standards and consumers everywhere (outside the U.S.) were feeling more confident about the future. On the other hand, congestion in cities was becoming a serious problem, water and air pollution were on the rise, and urbanization was compounding the development problems in large metropolitan areas.

This rapid, worldwide industrialization resulted in unprecedented use of petroleum products. By 2007, many scientists across the globe were raising environmental concerns and predicting dire consequences from the emergent threat of global warming. An evidence debate occupied a surprisingly large place in the daily news media — especially in Europe, but on the Internet, as well. From 2008 to 2010, a very strong consensus emerged in the scientific community (and in many influential parts of the global political community) that the world was on the brink of an environmental crisis because of high levels of CO_2 emissions. New measurement techniques (available only in recent years) demonstrated emphatically that global warming was occurring. It was further revealed that a small increase in the arctic temperature was leading to permafrost melting in the tundra. Scientists had known for over a decade that the melting tundra would release CO_2 and methane at levels so high as to seriously compound problems that had been created from industrialization.

The political reaction to the global environmental crisis in Europe was more decisive than in other parts of the world. Green parties became very powerful in the EU and their concerns quickly dominated political agendas. The Europeans had had CO_2 limits and CO_2 emissions credits systems for utilities since the 1990s. These were expanded dramatically. The Union also established stringent CO_2 emissions requirements on all emitting industries and vehicles. U.S. (and many emerging market) products that could not meet these standards were denied import licenses. Whereas, European products that could meet the new standards were marketed with growing success everywhere (even at somewhat higher prices).

Initially, U.S. responses were weak and vacillating owing primarily to the political reality that job protection had become the bottom line of economic and social policy. However, mounting evidence of an impending crisis over about two years created a mood swing in U.S. public attitudes (especially among the baby busters generation that had always been "green") that made a positive government response a necessity. U.S. industry driven by market concerns and export opportunities (the European CO_2 standard was fast becoming global) put even greater pressure on the government for decisive action in setting targets for emissions levels, supporting research, and ameliorating the social and economic consequences of attempting to meet the targets.

The most developed industrialized nations led several diplomatic moves for concerted global action. However, newly industrializing nations, led by China, were quite skeptical concerning the "Western" definition of the problem and were very reluctant to accept "Western" solutions. Some nations and many radical groups insisted that the CO_2 scare was a conspiracy of the industrialized nations to prevent the emergence of competitors.

Not satisfied with the global diplomatic malaise, the Europeans established very aggressive CO_2 emissions targets by 2012. The U.S. and Japan followed this lead and also established strict time tables for lowering CO_2 emissions. These targets and the regulations necessary to meet them became the progenitor for the development of new products and services across the globe. The rapid decrease demanded in CO_2 emissions also had the predictable effect of seriously reducing economic growth. The world slid into a deep recession for four to seven years (depending on the region).

By 2015 a complex mix of international cooperative and independent national actions was under way, within a constrained economic environment. These actions included: (1) national efforts in CO_2 control, including major regulations, use taxes on fuels, significant government sponsored research and development and incentives and penalties; (2) private research and development for solutions that gain market share in a CO_2 "limited" world; (3) international research and development efforts of corporations and nations; (4) international regulations agreements; (5)

national and international emissions targets; (6) the emergence of a World Environment Council that has been delegated authority to impose sanctions on offenders of emissions targets and regulations; (7) an international agreement with a phased set of targets to reduce world CO_2 emissions significantly by 2040.

The war on CO_2 emissions resulted in major economic dislocations, significant unemployment, and major increases in national debts as governments and corporations attempted to implement national and international strategies. As more restrictions were placed on the use of hydrocarbon fuels, the global recession deepened. Internal combustion automobiles became seldom used; factories curtailed operations unless they reduced CO_2 emissions; far fewer scheduled airliners flew and when they did it had to be with the most updated propulsion systems.

As more and more countries "came on board" with CO_2 limits (the physical evidence — including disruptive climate change — was simply becoming overwhelming), a fair adjudication of CO_2 emissions limits evolved from the European experience. Across the globe, CO_2 emissions "credits" were allocated to countries through multilateral negotiation based on a complex set of metrics including level of industrial development, size and density of population, number of forests, etc. Countries could buy and sell their credits — their "rights to pollute." As had happened in an earlier time in the 1980s and 1990s in the U.S. and Europe, a futures market in CO_2 credits emerged and global market forces pushed the sale, barter, and trade in credits as differing countries made choices about the approach they would take to meet global standards. Within countries the re-allocation of emissions credits was up to national governments. Some, as in Europe and many parts of Asia, chose strict highly regulated rationing programs. In the U.S. regulated set-asides for "nationally critical activities" were mixed with market mechanisms for private sector utilization of "emission rights."

The world has entered into a period of considerable social crisis with inevitable political and economic instability. There is continuing tension over hydrocarbon limits and credits and over national regulations and targets which often seem to favor local industries. Many nations have increased their defense budgets and strengthened their military forces to back up economic sanctions for states not cooperating with CO_2 reduction protocols and for the simple reason that the future has become highly uncertain. The dislocations associated with hydrocarbon limits have created civil unrest in many economies. Across the globe, these are very hard and acutely uncertain times.

Stagnant and recession economics has become the expected norm, globally. While oil supplies have remained stable and wholesale costs have not increased drastically (under complex multilateral treaties), end-user costs of products derived from oil have increased dramatically. Along with emissions credits, rationing of oil has become one other means of controlling CO_2 emissions. But this rationing has increased production costs as manufacturers invest heavily in alternative processes. Almost all nations are experiencing high unemployment, high inflation, high cost of capital, and major loss of net worth as stock markets stagnate and decline. Entire industry segments are now at risk, if they still exist. Corporations remain marginally competitive, if their markets are contiguous with their manufacturing location, but are in serious trouble if they must rely on long-range transportation.

Of all nations, the U.S. is among the worst off. Highly dependent on CO_2 emitting industries, our resources are stretched very thin as we attack the problems. The poor business judgments of earlier times, combined with governments that recklessly pursued social spending and higher public debt, left the U.S. with few options in these very hard times of restricted petroleum use, high unemployment, and interminable recession. Government debt is high and private corporations are stretched to the limit; yet vast resources are necessary to find both immediate and long-term relief of CO_2-related problems. Much of the U.S. population has shaken off its earlier lethargy; Americans generally rise to any challenge. Yet they find the most vigorous actions are centered in Europe, and many join those efforts through the Internet to make the most of their contributions to the global crisis.

Scenario Matrix - Environmentally Challenged

Scenarios / Drivers	Environmentally Challenged
World Economy and Market Environment	Global recession; entire industry segments at risk, some segments doing quite well; lots of environmental rules; high unemployment, high inflation, high cost of capital; stock markets have tumbled, bankruptcies high among energy dependent companies; financial stability imposed by international government coordination.
International Trade Environment	Intra-regional trade agreements prevail; cooperation within trading groups; global harmonization on environmental issues; landing rights and air space highly regulated and tied to emissions credits; environmental crisis forces international cooperation.
Political Instability	High political and economic instability; tension over hydrocarbon limits; trade tensions; civil instability in disrupted economies; the future contains very high levels of uncertainty; some rearmament (significant in some cases).
U. S. Military Requirements	Need to be able to threaten military sanctions for states not cooperating with CO_2 reduction protocols; cooperative actions; tendency to want to find non-military solutions to problems; DOD pursues hydrocarbon free power systems; moderate relatively high-tech world arms market.
Global Distribution of Power & Technology	Japan, China, US, and the European Union are key powers; key to global leadership is intellectual capital (especially in science and research) - European Union may be leader in global science and research activity.
Fuels & Fuel Sources	Stable oil supplies; wholesale cost is stable and maintained through negotiated international agreement on fair price for suppliers; rationing of CO_2 emission limits the use of hydrocarbon fuels; huge global effort searching for alternative fuels.
US Policy	Global environmental crisis results in bifurcated US policy responses including international scientific and environmental cooperation on the one hand and economic protectionism on the other; labor issues are highly politicized; in response to Europe's leadership in the reduction of CO_2 emissions, pragmatic US cooperation emerges; no tort reform (possible reversal of earlier reforms); very US centric and populist policies tempered by

Drivers \ Scenarios	Environmentally Challenged
	strong regional (NAFTA +) links; lenient anti-trust interpretation and enforcement to buttress US industries; mixed pressures on bankruptcy decisions related to environmental issues, job retention, and existence of national industries; very high deficit spending and national debt; low humanitarian consensus even regarding Latin America (just trade, not aid).
Corporate Structure and Operations	Corporations are competitive if their markets are contiguous with manufacturing location but are not competitive if they rely on long range transportation; there are incentives for the proliferation of technologies impacting products or services that reduce CO_2 emissions; developing countries leverage for requiring offsets is primarily limited to CO_2 emission credits; environmental regulations result in high production costs and high unemployment; US job retention policies burden firms with high labor costs; inter-firm alliances across geographic regions to optimize CO_2 credits.
Environment	High sensitivity to all pollution issues, but spending focus is on CO_2; long term effects of all products and activities are scrutinized; alternatives to hydro-carbons will have high threshold of acceptance.
Public Health	Carriers for many diseases have greater range with warmer climate; health problems from mass migration away from climate impacted areas; climatic shifts have huge impacts on public health globally; public health funding becomes serious competitor for public resources.
Public Attitude to Technology	Technology is the problem!; technology is the savior!; this debate may form the locus of political discourse; general anti-(new) technology bias.
Education	US public education is under-funded; the funding that is available flows into focused university research and global work in environmental science; independent and private funding is in applied science and technology; non-science education gets very little support; virtual education is finally being supported.
Geographic (Living) Dispersion	Bicycle world; huge disincentives to disperse; clustering near work/stores.
Communications and Information Technology	For about the first decade, strong global growth in communication, but US growth in communications and information systems tended to lag even as US companies

Scenarios / Drivers	Environmentally Challenged
	helped drive global development; in last years communication is leveraged in all ways possible to substitute for transportation; significant computer-controlled energy management.
Production Cost Performance	Cost of manufacturing goes up as policies to reduce CO_2 are implemented, driven by pollution credits; alternative fuels crucial; manufacturing location decisions must consider local pollution credits and transport costs.
Technology development and Application	Very unidimensional technology development - focused on detecting, modeling, and forecasting CO_2 plus energy conservation (demand and supply side), alternate fuels and sources of energy, technology to "absorb" carbon and store in benign forms; science is done globally and technology is done locally.
Time Poverty Leisure Time, Entertainment	Time poverty is not a serious problem for most people; very local leisure and entertainment; home entertainment is very important; town picnics, county fairs, and tree planting outings.
Global Transportation Infrastructure	Integrated infrastructure designed to reduce CO_2 emissions; regulated access to infrastructure; trend toward smart infrastructure (e.g., metered auto access to highways).
Safety and Security	Other issues overshadow safety; security threat (including data security) is moderate, disbursed, and comes from nations and groups dissatisfied with limits to growth and development.
Access to space	Very hard to justify unless attached to solution to CO_2 problem, such as space based sensors or energy source.

APPENDIX E

Workshop Agenda

> **Monday, September 30–Wednesday October 2, 1996**
> *Arnold and Mabel Beckman Center*
> *100 Academy Drive*
> *Irvine, California 92715*

Monday, September 30, 1996

OPENING SESSION

8:00 a.m. Welcome, Introductions, and Steering Committee Chair's Opening Remarks
>*William W. Hoover*

8:30 Remarks from the Associate Administrator for Aeronautics and Overview of the NASA Aeronautics Program
>*Robert Whitehead*

9:00 NASA Strategic Planning Process
>*Robert Pearce, NASA Office of Aeronautics*

9:30 Scenario Development Process
>*Charles Thomas, The Futures Group*

WORLD TEAM SESSIONS

10:30 World Teams 1 to 5 convene separately to begin work on refining the scenarios, determining needs and opportunities, and addressing the agenda book questions

PLENARY SESSION

5:15 p.m.   Preliminary Presentations by Team Leaders

Tuesday, October 1, 1996

PLENARY SESSION

8:00 a.m.   Discuss results of previous day and schedule for current day

WORLD TEAM SESSIONS

8:30   World Teams 1 to 5 convene separately to complete work

WORLD TEAM ROUND-ROBIN SESSIONS

3:30 p.m.   Team Leaders meet with each of the other world teams to facilitate feedback and additional input (rounds 1–4)

7:30   ASEB steering committee executive session

Wednesday, October 2, 1996

NRC PLENARY SESSION

8:00 a.m.   Discuss NASA's future role in aeronautics R&D

PLENARY SESSION

8:30   Group discussion of cross-cutting ideas

Thomas Sheridan

Appendix E

WORLD TEAM SESSIONS

9:30　　　Group closure on questions 1 to 6 and follow-up from round-robin sessions

- Discussion of technology implications - question 8
- Focus on high scoring opportunities and needs
- Discuss implications of lower- scoring opportunities and needs
- Question 7

PLENARY SESSION

2:15 p.m.　　Analysis of robust opportunities and needs
　　　　　　Discuss technology implications of opportunities and needs

3:15　　　Adjourn

3:30　　　Steering Committee Executive Session

6:00　　　Steering Committee Adjourns

APPENDIX F

Questions for World Team Sessions

(1) How does the consumer influence the marketplace around the world? What is the impact of lifestyle and cultural differences? Are there regional or national constraints or factors that influence consumer behavior? Within these considerations, how do consumers utilize global transportation? Have any trade-offs been made between the use of communications systems and the use of transportation systems? What does the aviation component look like? What does access to space look like?

(2) Discuss the nature of general business activities around the world. How are they servicing their customers? What are the constraints on their behavior? How do businesses utilize global transportation? Have any trade-offs been made between the use of communications systems and the use of transportation systems? What does the aviation component look like? What does the access to space component look like?

(3) Discuss the role of local, national, regional, and/or global governmental and regulatory authorities toward the transportation sector and the global aeronautics products and services industry.

(4) Summarize how the global civilian aeronautics products and service industry operates.

(5) What is the military security environment like? What is the impact of international or domestic terrorism? What is the role of the United States in dealing with international security? What is the role of the U.S. military? Describe the aeronautics component of that role. Describe the importance of access to space in that role.

(6) Given your answers to the above questions, what are the needs and opportunities for the U.S. aeronautics products and services industry? The needs and opportunities should recognize national goals such as maintaining the superiority of U.S. aeronautics products and services by enhancing performance, efficiency, affordability, and survivability; achieve an efficient,

safe, and affordable global air transportation system by improving capacity and efficiency and safety and security; ensuring the long-term environmental compatability of the aviation system. What are the technological implications of these needs and opportunities?

(7) Look beyond the time horizon of your world and speculate about possible innovations or breakthrough initiatives that you envision could change the nature of your world. Be creative with this question but stay true to a reasonable interpretation of how the future of your world might develop. What are the major obstacles that need to be overcome? Which of these can be addressed by advances in aeronautics or access to space? Is there an overriding need that cannot be ignored?

(8) Describe the research and development activities that could address the technological implications of your world's global needs and opportunities. Prioritize these research and development activities. How is this research and development organized and conducted? What respective roles should the U.S. government, academia, and industry play? What role should NASA play?

APPENDIX G

Bibliography

Clinton, W.J. and A. Gore, Jr. 1993. Technology for America's Economic Growth, a New Direction to Build Economic Strength. Washington, D.C.: Office of the President.

Coates, J.F. 1996. The Highly Probable Future: 83 Assumptions About The Year 2025. Bethesda, Md: The World Future Society.

Congressional Budget Office. 1996. Deficit Reduction Options: Eliminate NASA's Support for Producers of Commercial Airplanes. Washington, D.C.: U.S. Government Printing Office.

Cornish, E. 1996. The Cyber Future: 92 Ways Our Lives Will Change by the Year 2025. Bethesda, Md: The World Future Society.

Council on Competitiveness. 1996. Endless Frontier, Limited Resources: U.S. R&D Policy for Competitiveness. Washington, D.C.: Council on Competitiveness.

Federal Aviation Administration. 1996. FAA Long-Range Aviation Forecasts Fiscal Years 2007–2020. Office of Aviation Policy and Plans. FAA-APO-96-5. Washington, D.C.: Federal Aviation Administration.

Gellman Research Associates, Inc. 1992. Economic Analysis of Aeronautical Research and Technology—An Update. Unpublished Report. Gellman Research Associates, Inc., Jenkintown, Pa.

Langford, J.S., III. 1989. The NASA Experience in Aeronautical R&D: Three Case Studies with Analysis. Washington, D.C.: Institute for Defense Analyses.

Mowry, D.C. 1987. Alliance Politics and Economics: Multinational Joint Ventures in Commercial Aircraft. Cambridge, Mass.: American Enterprise Institute/Ballinger.

National Academy of Engineering. 1993. The Future of Aerospace: Proceedings of a Symposium held in Honor of Alexander H. Flax, Home Secretary, National Academy of Engineering. Washington, D.C.: National Academy Press.

National Academy of Engineering. 1993. Mastering a New Role: Shaping Technology Policy for National Economic Performance. Series on "Prospering in a Global Economy." Washington, D.C.: National Academy Press.

National Aeronautics and Space Administration. 1995. Achieving Aeronautics Leadership: Aeronautics Strategic Enterprise Plan, 1995–2000. Washington, D.C.: NASA.

National Aeronautics and Space Administration. 1994–1995. AGATE Program Information Binder. NASA Langley Research Center, Langley, Va. (Information on Developing Technologies for Business and Personal Transportation Aircraft.)

National Research Council. 1984. Aeronautics Technology Possibilities for 2000: Report of a Workshop. Aeronautics and Space Engineering Board. Washington, D.C.: National Academy Press.

National Research Council. 1985. Aeronautical Technology 2000: A Projection of Advanced Vehicle Concepts. Aeronautics and Space Engineering Board. Washington, D.C.: National Academy Press.

National Research Council. 1985. The Competitive Status of the U.S. Civil Aviation Manufacturing Industry: A Study of the Influences of Technology in Determining International Industrial Competitive Advantage. Committee on Technology and International Economic and Trade Issues, National Academy of Engineering. Washington, D.C.: National Academy Press.

National Research Council. 1992. Aeronautical Technologies for the Twenty-First Century, Executive Summary. Committee on Aeronautical Technologies, Aeronautics and Space Engineering Board. Washington, D.C.: National Academy Press.

National Research Council. 1992. Trends and Issues in International Aviation. Transportation Research Circular 393, Transportation Research Board. Washington, D.C.: National Academy Press.

National Research Council. 1994. Airport and Airspace Planning and Operations. Transportation Research Record 1461, Transportation Research Board. Washington, D.C.: National Academy Press.

National Research Council. 1994. Assessing the National Plan for Aeronautical Ground Test Facilities. Aeronautics and Space Engineering Board. Washington, D.C.: National Academy Press.

National Research Council. 1994. Future Aviation Activities: Eighth International Workshop. Transportation Research Circular 425, Transportation Research Board. Washington, D.C.: National Academy Press.

National Research Council. 1994. High-Stakes Aviation: United States - Japan Technology Linkages in Transport Aircraft. Office of International Affairs. Washington, D.C.: National Academy Press.

National Research Council. 1995. Airport and Air Transportation Issues. Transportation Research Record 1506, Transportation Research Board. Washington, D.C.: National Academy Press.

Office of Industry Assessment. 1984. A Competitive Assessment of the U.S. Civil Aircraft Industry. Industry Analysis Division, International Trade Administration. Washington, D.C.: U.S. Department of Commerce.

Office of Science and Technology Policy. 1982. Aeronautical Research and Technology Policy. Vol. I: Summary Report. Washington, D.C.: Executive Office of the President.

Office of Science and Technology Policy. 1982. Aeronautical Research and Technology Policy. Vol. II: Final Report. Washington, D.C.: Executive Office of the President.

Office of Science and Technology Policy. 1987. National Aeronautics R&D Goals: Agenda for Achievement. Washington, D.C.: Executive Office of the President

Office of Technology Assessment. 1991. Competing Economies: America, Europe and the Pacific Rim. Summary. OTE-ITE-499. Washington, D.C.: Office of Technology Assessment.

Office of Technology Assessment. 1994. Federal Research and Technology for Aviation. Washington, D.C.: Office of Technology Assessment.

Tassey, G. 1995. Technology and Economic Growth: Implications for Federal Policy. Program Office, National Institute of Standards and Technology. Gaithersburg, Md: U.S. Department of Commerce.

Tyson, L.D. 1992. Who's Bashing Whom? Trade Conflict in High-Technology Industries. Washington, D.C.: Institute for International Economics.

U.S. Air Force Scientific Advisory Board. 1995. New World Vistas: Air and Space Power for the 21st Century, Aircraft and Propulsion Volume. Washington, D.C.: U.S. Air Force Scientific Advisory Board.

U.S. International Trade Commission. 1993. Global Competitiveness of U.S. Advanced-Technology Manufacturing Industries: Large Civil Aircraft. Pub. No. 2667. Washington, D.C.: U.S. International Trade Commission.

Volpe National Transportation Systems Center. 1995. Symposium on Challenges and Opportunities for Global Transportation in the 21st Century. Transportation Strategic Planning and Analysis Office. Washington, D.C.: U.S. Department of Transportation.

APPENDIX H

Scenario-Based Strategic Planning as Described by the Futures Group[1]

Charles M. Thomas
The Futures Group

Scenario-based planning is a technique for managing uncertainty not prediction. Organizations typically choose this planning technique when

- the ambiguity in their operating environment is high
- the pace of change and degree of turmoil is accelerating
- their planning horizon tends to stretch out to ten years or more

When scenarios are used for planning, the intent is not to predict what the market will be and then build a master plan, but rather to ask what the future might hold and then identify the actions that can be taken today that will work no matter how the future turns out. As a result, the technique tends to rely more on expert judgment and less on quantitative forecasts such as market size or share. Although the payoff from the use of scenarios can be quite high, their use and development require intensive and highly customized efforts.

To develop scenarios, an organization must first cast a wide net so as to identify the many potential issues, trends, and factors in the operating environment (past, present, and future) that can have an impact on the organization. These issues and trends are called mission drivers or business drivers, and whether the objective is to explore organizational portfolio strategy or a technology investment strategy, this initial research and brainstorming should not be restricted.

With the drivers nominated, the next step is to reduce the breadth of the examination and bring some rigor and systematic analysis to categorizing the business drivers. This involves clustering and synthesizing all the drivers (typically as many as 150 to 200) into the macrolevel issues that define the overall decision planning environment. These macrolevel clusters become the fundamental assumptions of the planning effort, the base assumptions upon which the scenarios are built. These macrolevel assumptions are called "dimensions." They define the boundaries (dimensions) of the scenario planning space.

---

[1] This appendix has been edited for style and consistency.

To understand this better, one can take a visual image by thinking of a cube as "future space." Somewhere inside that cube is where an organization's future lies. That cube represents the planning space. The first task is to make sure that the cube encompasses all the issues relevant to the business. Everything that is important to consider must be "inside" that cube—those are the mission drivers. To define the dimensions, one describes the boundaries (the dimensions) of that cube or planning space. By defining the dimensions carefully, one sets the boundary conditions of the planning environment. The scenarios used are chosen from the inside of that cube, strategically selected to cover the range of threats and opportunities resident in the planning space that is defined with the dimensions. NASA, outside experts, and the Aeronautics and Space Engineering Board (ASEB) steering committee worked jointly to nominate the mission drivers, select the dimensions, and chose the scenarios to be studied in detail.

The scenario space that emerged from the dimensions chosen are shown in Figure H-1 (the dimensions are arrayed along the top). The scenarios chosen for development are highlighted.

U.S. Economic Competitiveness		Worldwide Demand for Aeronautics Products & Services		Threats to Global Security and/or Quality of Life		Global Trend in Government Participation in Society		
Strong	Weak	High Growth	Low Growth	High	Low	Low	High	World Name
X		X		X		X		Best for Aviation
X		X		X			X	Gun n' Butter
X		X			X	X		Pushing the Envelope
X		X			X		X	Jolly Green Giant
X			X	X		X		Weak for Aviation (I)
X			X	X			X	Grounded
X			X		X	X		Weak for Aviation (II)
X			X		X		X	Green Pastures
	X	X		X		X		Exhausted Super Crop
	X	X		X			X	Regional Tensions
	X	X			X	X		Trading Places
	X	X			X		X	They Won
	X		X	X		X		Dark Ages
	X		X	X			X	Environmentally Challenged
	X		X		X	X		Return to the Dark Ages
	X		X		X		X	Worst for Aviation

FIGURE H-1: NASA global scenarios—scenario space matrix.

The Futures Group, Science Applications International Corporation, and NASA spent approximately eight weeks developing the scenario details (carefully following the advice of the steering committee concerning the intent of the scenarios chosen) and producing the narratives. The scenario documents contained narrative future histories, scenario contingent forecasts of key time series data, and detailed examination of the end states in 2015–2020.

Scenario planning is a form of due diligence in planning. The organization is asked to "live" in each of the scenarios (in a sense, to insert their business judgment into each scenario) and ask "what should we do here to survive?" What are the customer needs *in my scenario*, what are the constraints on my behavior *in my scenario*, etc.? In this case, NASA and the ASEB steering committee divided themselves into "world teams" and investigated the needs for aeronautics products and services in *each* scenario independently. That independent work was then synthesized into a short list of needs and opportunities that were robust across all the scenarios. The output of a scenario planning exercise is a robust core set of product ideas, services, or organizational strategies that are viable no matter how the future actually evolves. Figure H-2 illustrates that process.

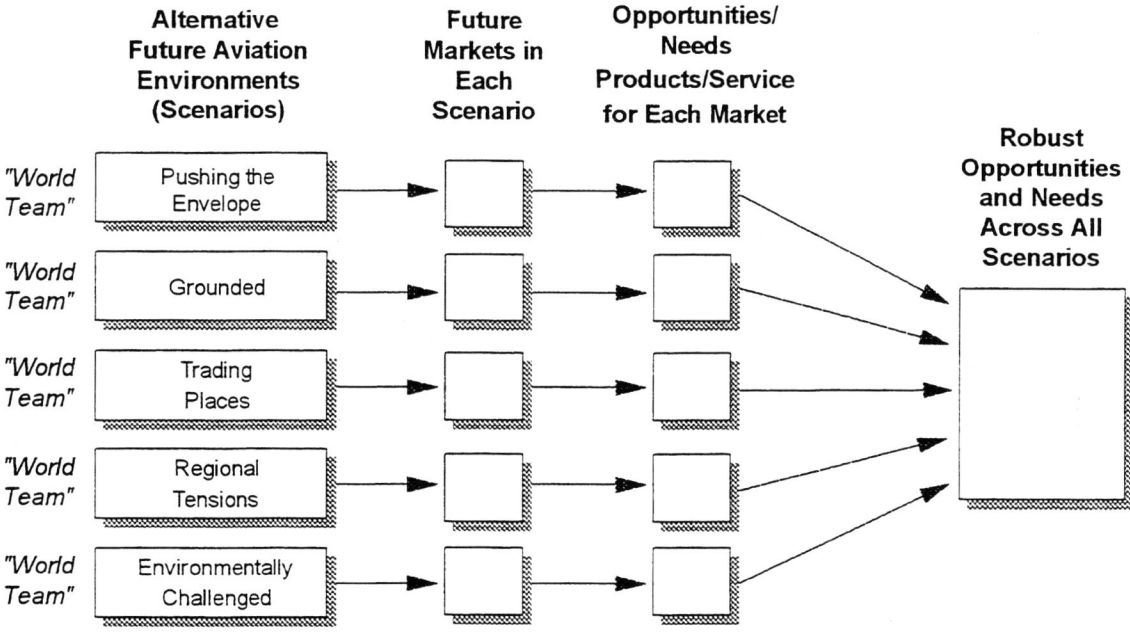

FIGURE H-2 Organizational strategies.

Scenario-based planning is a powerful tool to manage the uncertainties that surround important organizational or technology investment decisions. The scenarios provide decision makers with a window on potential future market needs. Scenarios provide a tool for a systematic and innovative consideration of new customer needs and a method for evaluating the long-term viability of current organizational structures, product ideas, or technology developments. Scenarios help set *future market-driven priorities* for technology investment plans so that they are more than just lengthy wish lists.